Beyond Trachtenberg

Beyond Trachtenberg

David Herrera Pérez

ISBN: 9798352163047

Contact: micifut@hotmail.com
URL of the author: https://amazon.com/author/davidhperez

Table of Contents

Preface

This is a self-help book for those who at some point in their lives lost their self-confidence; for those who believe they are incapable of solving any operation without a calculator; in short, for those who don't trust their own brain. The book explores non-traditional ways of calculating inspired by the knowledge of the Russian Jakow Trachtenberg, who, a prisoner in a Nazi concentration camp, knew how to maintain his sanity by devising how to operate with numbers only with his mind.

The reader will be surprised by methods excluded from traditional school learning, opening his/her mind to new possibilities of thought that prioritize the use of only addition and subtraction; he/she will check for himself/herself the logic inherent in the calculations and will be seduced by the algorithms used, which will inevitably make him/her use his/her brain.

The first reading of the book should be done from the first page to the last; the absence of indexes is deliberate, since a random reading of the chapters could discourage the reader, who, not being familiar with the new concepts, would experience some difficulty and confusion.

The book has a lot of useful information related to calculation, from tricks to multiply by the first natural numbers without the need of a calculator to methods to raise a number to any other, or to calculate by hand the nth root of a number or its logarithm, passing through uncommon curious ways of multiplying and dividing, the detection of errors in the operations, the decomposition in factors of the first 1000 numbers (highlighting among them those that are prime), the simplification of fractions, rules to determine the significant figures of a number and its rounding and a cursory look at the ABN method that complements traditional education.

Chapter 1

The inner magic of numbers

Some numbers seem to have in their essence an inner magic that they spread when we operate with them. This defining characteristic allows us to predict the result of the arithmetic operation in which they participate by following some simple rules that allow us to dispense with the use of huge amounts of paper and multiple pencils when facing the challenge of multiplying two numbers with many figures. In particular, Trachtenberg explored the basic tricks associated with **11**, **12**, **6**, **7**, **5**, **9**, **8**, **4**, **3**, **2**, **1**, and **0**. Some are obvious, but others... not so much.

All the numbers with which we are going to operate are in *base* 10 (the ones usually we find in daily life). In ***base* 10**, nothing is represented as **0**, one as **1**, $(1+1)$ as **2**, $(2+1)$ as **3**, $(3+1)$ as **4**, $(4+1)$ as **5**, $(5+1)$ as **6**, $(6+1)$ as **7**, $(7+1)$ as **8**, and $(8+1)$ as **9**; but in the representation of $(9+1)$ two numbers (10) are needed since all possible figures have been used and a new repetition cycle must be started (11, 12, 13, 14, 15, 16, 17, 18, 19) and then another (22, 23, 24, 25, 26, 27, 28, 29) and another... until the available spellings run out again and an additional figure is required on the left: $(99+1)$ is represented as 100. This subtle math trick is called **weighting**, and it makes calculations a lot easier. The position of *each digit of the number is intrinsically associated with a value* that (in *base* 10) corresponds to the logically expected one: *ten raised to the position of the digit within the number considered* (the positions are numbered from right to left, starting with 0). To clarify all this, let's consider the number **123**; **1** is at *position* 2 (10 squared is 100) so the value it contributes is 100 (1 times 100), **2** is at *position* 1 (10 to the power of 1 is 10) giving a value of 20 (2 times 10) y **3** is at *position* 0 (10 to the power of 0 is 1), determining the value 3 (3 times 1 is 3): their joint sum (100 plus 20 plus 3) constitutes the number 123. It is vitally important to bear in mind that when adding two numbers, one must always operate with the figures of equal weight (the units of one number with the units of the other, the tens of one with the tens of the other, the hundreds of one with the hundreds of the other... and so on), and so on starting with the least weight (those furthest to the right) and working towards the left until the available figures are exhausted. In each of these sums, if its value produces a two-digit number (that is, it exceeds 9), there is a carry (the tens of that sum) that must be dragged to the left until the next sum; e. g. adding 63 to 957 results in 1020 in these steps:

- 7 plus 3 equals 10 (**0** in the units and 1 *carry*).
- 5 plus 6 plus 1 *carry* equals 12 (**2** in the tens and 1 *carry*).
- 9 plus 0 plus 1 *carry* equals 10 (**0** in the hundreds

and 1 *carry* that is transferred to the thousands to be added there with 0, generating the digit **1**).

Multiplication by 11

We are now ready to investigate what the result of multiplying any number by 11. We will use 123 as the victim. The number 11 is 10 plus 1, so you have to multiply 123 by 1, 123 by 10 and add both results. Traditionally this is expressed like this:

0	**1**	**2**	**3**
	×	**1**	**1**
	1	2	3
1	2	3	
1	**3**	**5**	**3**

The number **123** is **100** (1 times 100 or $1 \cdot 10^2$) plus **20** (2 times 10 or $2 \cdot 10^1$) plus **3** (3 times 1 or $3 \cdot 10^0$) and **11** is **1** plus **10** (1 times 10 or 10^1), so:

$$
\begin{aligned}
\mathbf{123} \cdot \mathbf{11} = (100 + 20 + 3) \cdot (1 + 10) \\
&= (\mathbf{1} \cdot 10^2 + \mathbf{2} \cdot 10^1 + \mathbf{3} \cdot 10^0) \cdot 1 \\
&+ (\mathbf{1} \cdot 10^2 + \mathbf{2} \cdot 10^1 + \mathbf{3} \cdot 10^0) \cdot 10^1 \\
&= (\mathbf{1} \cdot 10^2 + \mathbf{2} \cdot 10^1 + \mathbf{3} \cdot 10^0) \\
&+ (\mathbf{1} \cdot 10^3 + \mathbf{2} \cdot 10^2 + \mathbf{3} \cdot 10^1)
\end{aligned}
$$

and grouping the terms of the same power of 10,

$$
\begin{aligned}
\mathbf{123} \cdot \mathbf{11} &= \mathbf{1} \cdot 10^3 + (\mathbf{1} + \mathbf{2}) \cdot 10^2 + (\mathbf{2} + \mathbf{3}) \cdot 10^1 + \mathbf{3} \cdot 10^0 \\
&= \mathbf{1} \cdot 10^3 + \mathbf{3} \cdot 10^2 + \mathbf{5} \cdot 10^1 + \mathbf{3} \cdot 10^0 = \mathbf{1353}
\end{aligned}
$$

An interesting rule can be inferred from here:

Each figure of the result can be obtained from the original number (to which we add one more figure to its left of null value for the convenience of the algorithm) without more than ***adding to the digit of its position the one that precedes it*** *(just the one on its right or 0 if it does not exist)* ***and the immediately preceding carry*** *(if applicable).*

In the example, the original number is 0123, so the process to follow is: 3 plus 0 is **3** (the units); 2 plus 3 equals **5** (the tens); 1 plus 2 is **3** (the hundreds); and 0 plus 1 is **1** (the thousands), which results in **1353**.

An example with carry is 178 (consider 0178 to know when to stop) times 11 whose result is **1958**: the ones are **8** (8 plus 0), the tens are **5** (7 and 8 are **15**, that is: **5** and 1 *carry*), the hundreds are **9** (1 plus 7 plus 1 *carry*) and the thousands are **1** (0 plus 1).

Multiplication by 12

Now we are going to investigate the result of multiplying any number by 12 using 123 as a victim. The number 12 is 10 plus 2, so you have to multiply 123 by 2, 123 by 10 and add both results:

0	**1**	**2**	**3**
	×	**1**	**2**
	2	4	6
1	2	3	
1	**4**	**7**	**6**

The number **123** is **100** (1 times 100 or $1 \cdot 10^2$) plus **20** (2 times 10 or $2 \cdot 10^1$) plus **3** (3 times 1 or $3 \cdot 10^0$) y **12** is

2 plus **10** (1 times 10 or 10^1), so:

$$\begin{aligned}\mathbf{123}\cdot\mathbf{11} &= (100+20+3)\cdot(2+10)\\ &= (\mathbf{1}\cdot 10^2+\mathbf{2}\cdot 10^1+\mathbf{3}\cdot 10^0)\cdot 2\\ &+ (\mathbf{1}\cdot 10^2+\mathbf{2}\cdot 10^1+\mathbf{3}\cdot 10^0)\cdot 10^1\\ &= (2\cdot\mathbf{1}\cdot 10^2+2\cdot\mathbf{2}\cdot 10^1+2\cdot\mathbf{3}\cdot 10^0)\\ &+ (\mathbf{1}\cdot 10^3+\mathbf{2}\cdot 10^2+\mathbf{3}\cdot 10^1)\end{aligned}$$

and grouping the terms of the same power of 10,

$$\begin{aligned}\mathbf{123}\cdot\mathbf{11} = \mathbf{1}\cdot 10^3+(2\cdot\mathbf{1}+\mathbf{2})\cdot 10^2+(2\cdot\mathbf{2}+\mathbf{3})\cdot 10^1+2\\ \cdot\mathbf{3}\cdot 10^0 = \mathbf{1}\cdot 10^3+\mathbf{4}\cdot 10^2+\mathbf{7}\cdot 10^1+\mathbf{6}\cdot 10^0\\ = \mathbf{1476}\end{aligned}$$

From where the following rule can be inferred:

Each figure of the result can be obtained from the original number (to which we add one more figure to its left of null value for the convenience of the algorithm) without more than ***adding to the DOUBLE of the digit of its position the one that precedes it*** *(the one on its right or 0 if it does not exist) and the* ***immediately previous carry*** *(if it has been produced).*

In the example, the original number is 0123 and the calculation is as follows: 3 times 2 is 6, plus 0 is **6** (the units); 2 times 2 is 4, plus 3 is **7** (the tens); 1 times 2 is 2, plus 2 is **4** (the hundreds); 0 times 2 is 0, plus 1 is **1** (the thousands), which results in **1476.**

The 178 by 12 product is an example with carry; the result is **2136**; the **units are 6** (8 times 2 plus 0 are 16, that is, **6** units and 1 *carry*), the **tens are 3** (7 times 2 plus 8 plus 1 *carry* is 23, that is: **3** plus 2 *carry*), the **hundred is 1** (the double of 1 plus 7 plus 2 *carry* is 11: the **1** on the right is the hundreds of the result, the one on the left determines 1 *of carry*) and the **thousands are 2** (2 times 0 is 0, plus 1 of the figure to its right, plus 1 corresponding to the immediate previous *carry*).

Multiplication by 6

Let's see what happens when we multiply any number by 6. As usual, we will use 123 as a victim:

0	**1**	**2**	**3**
		×	**6**
	7	**3**	**8**

Trachtenberg knew how to infer a rule by noting that 6 is 1 plus 5 (10 divided by 2).

The number **123** is **100** (1 times 100) plus **20** (2 times 10) plus **3**; **multiplying it by 6 is** multiplying **100** by 10, **20** by 10 and **3** by 10, dividing each result by 2, finding its sum and adding it to 123:

$$\mathbf{123} \cdot \mathbf{6} = 123 \cdot (1+5) = 123 \cdot \left(1 + \frac{10}{2}\right) =$$
$$= (100 + 20 + 3) \cdot \left(1 + \frac{10}{2}\right)$$
$$= (\mathbf{1} \cdot 10^2 + \mathbf{2} \cdot 10^1 + \mathbf{3} \cdot 10^0) \cdot \left(1 + \frac{10}{2}\right)$$

therefore,

$$\mathbf{123} \cdot \mathbf{6} = (\mathbf{1} \cdot 10^2 + \mathbf{2} \cdot 10^1 + \mathbf{3} \cdot 10^0) \cdot 1$$
$$+ (\mathbf{1} \cdot 10^2 + \mathbf{2} \cdot 10^1 + \mathbf{3} \cdot 10^0) \cdot \frac{10}{2}$$
$$= (\mathbf{1} \cdot 10^2 + \mathbf{2} \cdot 10^1 + \mathbf{3} \cdot 10^0) \cdot 1$$
$$+ \left(\frac{\mathbf{1}}{2} \cdot 10^3 + \frac{\mathbf{2}}{2} \cdot 10^2 + \frac{\mathbf{3}}{2} \cdot 10^1\right)$$

and grouping the terms of the same power of 10,

$$\mathbf{123} \cdot \mathbf{6} = \frac{\mathbf{1}}{2} \cdot 10^3 + \left(\mathbf{1} + \frac{\mathbf{2}}{2}\right) \cdot 10^2 + \left(\mathbf{2} + \frac{\mathbf{3}}{2}\right) \cdot 10^1 + \mathbf{3} \cdot 10^0$$

If the digit that is divided by 2 is odd, it is necessary to add 0.5 to it (the fractional part that is generated), which is equivalent to adding 5 to the immediate term of less weight. In this case 1 and 3 are odd, so their corresponding fractions are replaced by the quotient when dividing by 2 and the fractional part (5) is transferred to the adjacent term with less weight, so the previous expression becomes:

$$\begin{aligned}\mathbf{123}\cdot\mathbf{6} &= 0\cdot 10^3+(\mathbf{1}+1+5)\cdot 10^2+(\mathbf{2}+1)\cdot 10^1\\ &\quad+(\mathbf{3}+5)\cdot 10^0=\mathbf{7}\cdot 10^2+\mathbf{3}\cdot 10^1+\mathbf{8}\cdot 10^0\\ &= \mathbf{738}\end{aligned}$$

that is, the following rule is true:

Each figure of the result can be obtained from the original number (to which we add one more figure to its left of null value) without more than **adding** ***5** to the digit of its position (**only if it is odd**) plus the **INTEGER QUOTIENT** (without rounding) **that results from dividing by 2 the one that precedes it** (which is the one on its right or 0 if it does not exist) plus **the immediately previous carry** (if applicable).*

Comparing this rule with the result of the operation we see that: the **8** de of the units comes from the sum 3 (odd) plus 5 plus 0; the **3** of the tens is obtained by adding 2 (even) plus 1 (the integer part of dividing the number in the preceding position by 2, that is, 3 divided by 2); the **7** of the hundreds can be calculated as 1 (odd) plus 5 plus 1 (the integer part of 2 divided by 2); the thousands are 0 because to 0 (which we consider even) we would have to add the integer quotient of 1 between 2, which is 0.

The product 178 times 6 is an example with carry that results in **1068**: the **units are 8** (8 plus 0); the **tens are 6** (since **7** is odd, in addition to **4** —8 divided by 2— you have to add **5**, which results in **16**, that is: 6 and 1 *carry*), the **hundred is 0** (**1** plus **5** plus **3** —the integer part of 7 divided by 2— plus 1 *carry* add up to **10**, that is: 0 and 1 *carry*) and the **thousand is 1** (**0** —the additional figure added to the left of 178 for convenience of calculations plus **0** —the integer part of 1 divided by 2— plus 1 *carry*).

Multiplication by 7

What happens when you multiply any number by 7? As usual, we will use 123 as a victim. We have to add **1** + **20** (7 times 3 equals 21) plus **40** + **100** (7 times 20 equals 140) plus **700** (7 times 100), that is: the ones digit is **1**, the tens are **6** (**2**0 plus **4**0) and the hundreds **8** (**1**00 plus **7**00).

Traditionally the operation is like this: 7 times 3 equals 21 (I write a **1** in the units of the result and set aside 2); 7 times 2 is 14, which plus 2 *carry* makes 16 (I put **6** in the tens of the result and take 1 *carry*); finally, I multiply 7 by 1 (which is 7) and add the immediately previous carry (I had 1 *carry*) giving a total of **8** for the hundreds of the result. This whole process is usually summed up like this:

0	**1**	**2**	**3**
		×	**7**
	8	**6**	**1**

Trachtenberg knew how to infer a rule by noting that 7 is 2 plus 5 (10 divided by 2). The number **123** is **100** (1 times 100) plus **20** (2 times 10) plus **3**; **multiplying it by 7 is** multiplying 100 by 10, 20 by 10 and 3 by 10, then dividing each sum by 2, and adding the double of 123 to the total of the sum:

$$\begin{aligned}\mathbf{123}\cdot\mathbf{7} &= 123\cdot(2+5) = (100+20+3)\cdot\left(2+\frac{10}{2}\right)\\ &= (\mathbf{1}\cdot 10^2+\mathbf{2}\cdot 10^1+\mathbf{3}\cdot 10^0)\cdot\left(2+\frac{10}{2}\right)\end{aligned}$$

Operating and grouping the terms in powers of 10,

$$\begin{aligned}\mathbf{123}\cdot\mathbf{7} = {} & (\mathbf{1}\cdot 10^2+\mathbf{2}\cdot 10^1+\mathbf{3}\cdot 10^0)\cdot 2\\ & +(\mathbf{1}\cdot 10^2+\mathbf{2}\cdot 10^1+\mathbf{3}\cdot 10^0)\cdot\frac{10}{2}\\ & =(2\cdot\mathbf{1}\cdot 10^2+2\cdot\mathbf{2}\cdot 10^1+2\cdot\mathbf{3}\cdot 10^0)\\ & +\left(\frac{\mathbf{1}}{2}\cdot 10^3+\frac{\mathbf{2}}{2}\cdot 10^2+\frac{\mathbf{3}}{2}\cdot 10^1\right)\\ & =\frac{\mathbf{1}}{2}\cdot 10^3+\left(2\cdot\mathbf{1}+\frac{\mathbf{2}}{2}\right)\cdot 10^2+\left(2\cdot\mathbf{2}+\frac{\mathbf{3}}{2}\right)\cdot 10^1\\ & +2\cdot\mathbf{3}\cdot 10^0\end{aligned}$$

If the digit that is divided by 2 is odd, it is necessary to add 0.5 to it (the fractional part that is generated), which is equivalent to adding 5 to the immediate term of less weight. In this case 1 and 3 are odd, so that their corresponding fractions are replaced by the quotient when dividing by 2 and the fractional part is transferred to the adjacent term with less weight (adding 5), so the previous expression becomes:

$$\mathbf{123}\cdot\mathbf{7} = 0\cdot 10^3 + (2\cdot\mathbf{1}+5+1)\cdot 10^2 + (2\cdot\mathbf{2}+1)\cdot 10^1 + (2\cdot\mathbf{3}+5)\cdot 10^0 = \mathbf{8}\cdot 10^2 + (\mathbf{5}+\underline{\mathbf{1}})\cdot 10^1 + \mathbf{1}\cdot 10^0 = \mathbf{861}$$

where the tens generated by $2\cdot 3+5$ constitute the carry ($\underline{1}$) that is transmitted to the immediately higher weighting digit; that is, the following rule is fulfilled:

Each figure of the result can be obtained from the original number (to which we add one more figure to its left of null value) without more than adding ***to the DOUBLE of the digit*** *of its position:* ***5*** *(only if it is odd) plus the* ***INTEGER QUOTIENT that results from dividing by 2 the one that precedes it*** *(which is the one on its right or 0 if it does not exist) plus the* ***immediately preceding carry*** *(in case of occurrence).*

If we look at the result of operation (861), the **1** of the units comes from the ***double of* 3** (odd) plus **5** plus **0**, which gives a total of **11**, that is: **1** and 1 *carry*; the tens can be obtained by adding the ***double of* 2** (even) plus **1** *(the integer part of dividing the digit of the preceding position by 2, that is, 3 by 2)* plus **1** *carry*, which resulting in **6**; the hundreds can be obtained in a similar way by adding the ***double of* 1** (because 1 is odd) plus **5** plus **1** (the integer part of 2 divided by 2) resulting in **8**; the thousands are 0 because to the 0 (even) we would have to add the integer quotient of 1 between 2, which is 0.

An *example with carry* is 0178 · 7 whose result is **1246**: the **units are** 8 times 2 plus 0, which gives **16** (**6** plus 1 *carry*); the **tens are 4** —to the ***double of* 7** we must add **5** (because 7 is odd) plus **4** (the quotient of 8 between 2) plus **1** *carry*, which results in **24** (4 and 2 *carry*)—, the **hundred is 2** —the ***double of* 1** plus **5** (1 is odd) plus **3** (the integer part of 7 divided by 2) plus **2** *carry*, which adds up to **12** (2 and 1 *carry*)— and the

thousands are 1 —the **0** corresponding to the double of the additional digit added for convenience of calculation, plus **0** of the integer part of the result of the operation 1 divided by 2, plus **1** *carry*—.

Multiplication by 5

Having the number 5 as a multiplier is quite friendly. Traditionally the operation is indicated like this:

0	**1**	**2**	**3**
		×	**5**
	6	**1**	**5**

We have used as multiplicand **123**, which is **100** (1 times 100) plus **20** (2 times 10) plus **3**; **multiplying it by 5 is** multiplying 100 by 10, 20 by 10 and 3 by 10, dividing each addend by 2 and adding the results:

$$\mathbf{123} \cdot \mathbf{5} = 123 \cdot \frac{10}{2} = (100 + 20 + 3) \cdot \frac{10}{2}$$
$$= (\mathbf{1} \cdot 10^2 + \mathbf{2} \cdot 10^1 + \mathbf{3} \cdot 10^0) \cdot \frac{10}{2}$$
$$= \left(\frac{\mathbf{1}}{2} \cdot 10^3 + \frac{\mathbf{2}}{2} \cdot 10^2 + \frac{\mathbf{3}}{2} \cdot 10^1\right)$$

If the digit that is divided by 2 is odd, it is necessary to add 0.5 (the fractional part that is generated), which is equivalent to adding 5 to the immediate term of less weight. In this case 1 and 3 are odd, so their corresponding fractions are replaced by the quotient when dividing by 2 and the fractional part (5) is transferred to the adjacent term with less weight, so the previous expression becomes:

$$\mathbf{123} \cdot \mathbf{5} = 0 \cdot 10^3 + (1 + \mathbf{5}) \cdot 10^2 + 1 \cdot 10^1 + (0 + \mathbf{5}) \cdot 10^0$$
$$= \mathbf{6} \cdot 10^2 + \mathbf{1} \cdot 10^1 + \mathbf{5} \cdot 10^0 = \mathbf{615}$$

that is, the following rule is fulfilled:

Each figure of the result can be obtained from the original number (to which we add one more figure to its left of null value) without more than **adding** ***5*** *(only if it is odd) to the* ***TRUNCATED INTEGER QUOTIENT*** *(without rounding)* ***that results from dividing by 2 the one that precedes it*** *(which is the one on its right or 0 if it does not exist).*

The ***immediately preceding carry*** *never occurs because the maximum number generated by this* *algorithm for any number of the result is* ***9****, that is,* ***5*** *plus* ***4*** *(the integer quotient of dividing 9 by 2).*

Let's compare the result of the operation and its possible relationship with each digit of the multiplicand:

The **5** of the ones is **0** plus **5** (the ones digit of the multiplicand is 3 —odd); the **1** of the tens is the **integer quotient** (truncated) that results **when dividing 3 by 2** (it is not necessary to add 5, since the tens digit of the multiplicand is even); and the **6** of the hundreds comes from **1** (the *truncated* half of 2, the digit of the previous position) plus **5** (since the digit of the hundreds of the multiplicand is 1, odd).

An example is 983 (we consider 0983 to facilitate the explanation) times 5 which results in **4915**:

The number 3 is odd, so you have to add **5** to **0** (the integer quotient that results from dividing 0 by 2) which results in a value of **5** for the **units** of the result; 8 is even, so the number that determines the **tens** is **1** (the integer quotient when dividing 3 by 2); the 9 of the multiplicand is odd, which makes the **hundreds** of the result **9** (the integer quotient when dividing 8 by 2, plus 5); finally, the 0 (which was added to facilitate the algorithm) is considered even, so the **thousands** digit is **4** (the integer quotient resulting from dividing the 9 of the multiplicand by 2).

Multiplication by 9

Having the number 9 as a multiplier may seem like a complicated task, but in truth it is not so much; Trachtenberg devised an algorithm that requires only addition and subtraction, the details of which are given below. As multiplicand we will use the handy 123. As usual, first of all we indicate the traditional way of operating:

0	**1**	**2**	**3**
		×	**9**
1	**1**	**0**	**7**

The number **123**, is **100** (1 times 100) plus **20** (2 times 10) plus **3**; **multiplying it by 9 is** multiplying 100 by 9,

20 by 9 and 3 by 9 and adding these results; but since 10 minus 1 is 9, we can say that:

$$\begin{aligned}\mathbf{123 \cdot 9} &= (100 + 20 + 3) \cdot (10 - 1) \\ &= (100 + 20 + 3) \cdot 10 - (100 + 20 + 3) \cdot 1 \\ &= 1000 + 200 + 30 - 100 - 20 - 3 \\ &= 1000 - 100 + 200 - 20 + 30 - 3\end{aligned}$$

Subtracting and adding 900, 90 and 9 (and grouping):

$$\begin{aligned}\mathbf{123 \cdot 9} &= 1000 - \mathbf{900} + \mathbf{900} - 100 + 200 - \mathbf{90} + \mathbf{90} - 20 \\ &\quad + 30 - \mathbf{9} + \mathbf{9} - 3 \\ &= 1000 + \mathbf{900} - 100 + 200 + \mathbf{90} - 20 + 30 \\ &\quad + \mathbf{9} - 3 - \mathbf{900} - \mathbf{90} - \mathbf{9} \\ &= 10^3 + (\mathbf{9} - 1 + 2) \cdot 10^2 + (\mathbf{9} - 2 + 3) \cdot 10^1 \\ &\quad + (\mathbf{9} - 3) \cdot 10^0 - (\mathbf{900} + \mathbf{90} + \mathbf{9})\end{aligned}$$

But since 999 (900 plus 90 plus 9) can be put as 1000 (10 cubed) minus 1:

$$\begin{aligned}\mathbf{123 \cdot 9} &= 10^3 + (9 - 1 + 2) \cdot 10^2 + (9 - 2 + 3) \cdot 10^1 \\ &\quad + (9 - 3) \cdot 10^0 - (\mathbf{10^3} - \mathbf{1}) \\ &= (1 - \mathbf{1}) \cdot 10^3 + (9 - 1 + 2) \cdot 10^2 \\ &\quad + (9 - 2 + 3) \cdot 10^1 + (9 - 3 + \mathbf{1}) \cdot 10^0 \\ &= (\mathbf{1} - 1) \cdot 10^3 + (9 - \mathbf{1} + \mathbf{2}) \cdot 10^2 \\ &\quad + (9 - \mathbf{2} + \mathbf{3}) \cdot 10^1 + (10 - \mathbf{3}) \cdot 10^0\end{aligned}$$

which allows us to say that when we multiply by 9, *each*

digit of the result can be obtained from the original number by adding and subtracting so that:

- *The figure located furthest to the right of the result is obtained by subtracting from 10 (**the complement of 10**) the units of the multiplicand.*
- *The remaining figures of the result (except the last) are calculated from right to left by adding to **the complement of** 9 **of the digit of the same position** of the multiplicand, **the one that precedes it** (that is, it is the one on its right) and the **carry** (if applicable).*
- *The leftmost digit of the result is calculated **by subtracting 1** from the **leftmost digit** of the multiplicand and **adding the carry** (if applicable).*

In our example:

$$\mathbf{123} \cdot \mathbf{9} = (1 - \mathbf{1}) \cdot 10^3 + (9 - \mathbf{1} + \mathbf{2}) \cdot 10^2 + (9 - \mathbf{2} + \mathbf{3}) \cdot 10^1 + (10 - \mathbf{3}) \cdot 10^0 = (0 + \mathbf{1}) \cdot 10^3 + (0 + \mathbf{1}) \cdot 10^2 + (0 + \mathbf{0}) \cdot 10^1 + 7 \cdot 10^0 = \mathbf{1} \cdot 10^3 + \mathbf{1} \cdot 10^2 + \mathbf{0} \cdot 10^1 + \mathbf{7} \cdot 10^0 = \mathbf{1107}$$

where each carry has been moved to the digit of the

immediate higher power; Let's see it narrated:

The **7 of the ones** is **10** minus **3**; The **0 in the tens** comes from **10** (**0** and 1 *carry*), that is, **3** (the immediately preceding digit) plus the *9's complement* of 2 (**9** minus **2**); the **1 of the hundreds** is 11 (**1** in the tens and 1 *of carry*), that is, **2** (the one immediately before) plus the *9's complement* of 1 (**9** minus **1**) plus **1** *carry*; and the **1 of the thousands** comes from **subtracting 1** from 1 (that of the multiplicand) and adding the carry (1 minus 1 plus 1 is 1).

One more example is 971 (consider 0971 to make the algorithm easier) times 9, which results in **8739**:

The **9 of the ones** is **10** minus **1**; the **3 in the tens** comes from **2** —the *9's complement* of 7 (subtract 7 from 9)— plus **1** (the immediately preceding digit); the **7 of the hundreds**, of **0** (the *9's complement* of 0, that is, 9 minus 9) plus **7** (the immediately preceding digit); and the **8 of the thousands** it's just **9** (the one immediately preceding the multiplicand **minus 1**) since there is no carry.

An example with two figures is 34 (consider 034 to facilitate the algorithm) times 9 whose result is **306**:

The **6 of the ones** is **10** minus **4**; the **0 of the tens** comes from **10** (0 and 1 *of carry*), that is, **6** —the *9's complement* of 3 (9 minus 3) plus **4** (the immediately preceding digit)—; and the **3 of the hundreds** comes from subtracting 1 from the 3 of the multiplicand and adding the carry.

Multiplication by 8

Having 8 as a multiplier is similar to multiplying by 9; the algorithm is almost the same, it only requires multiplying some numbers by 2 in addition to adding and subtracting; details are explored below. As multiplicand we will use, as always, 123. The traditional way of operating looks like this:

0	**1**	**2**	**3**
		×	**8**
0	**9**	**8**	**4**

The number **123**, is **100** (1 times 100) plus **20** (2 times 10) plus **3**; **multiplying it by 8 is** multiplying100 by 8, 20 by 8 and 3 by 8 (and adding them all up); but since 10 minus 2 is 8, we can say that:

$$\begin{aligned}\mathbf{123}\cdot\mathbf{8} = (100+20+3)\cdot(10-2) \\ &= (100+20+3)\cdot 10-(100+20+3)\cdot 2 \\ &= 1000+200+30-2\cdot 100-2\cdot 20-2\cdot 3 \\ &= 1000-2\cdot 100+200-2\cdot 20+30-2\cdot 3\end{aligned}$$

Subtracting and adding 1800, 180 and 18 (*twice* 900, 90 and 9, respectively) and grouping:

$$\begin{aligned}\mathbf{123}\cdot\mathbf{8} = 1000-\mathbf{1800}+\mathbf{1800}-2\cdot 100+200-\mathbf{180} \\ &+\mathbf{180}-2\cdot 20+30-\mathbf{18}+\mathbf{18}-2\cdot 3 \\ &= 10^3+\mathbf{18}\cdot 10^2-2\cdot 10^2+2\cdot 10^2+\mathbf{18}\cdot 10^1 \\ &-2\cdot 2\cdot 10^1+3\cdot 10^1+\mathbf{18}\cdot 10^0-2\cdot 3\cdot 10^0 \\ &-\mathbf{1800}-\mathbf{180}-\mathbf{18} \\ &= 10^3+(18-2\cdot\mathbf{1}+\mathbf{2})\cdot 10^2 \\ &+\ (18-2\cdot\mathbf{2}+\mathbf{3})\cdot 10^1+(18-2\cdot\mathbf{3})\cdot 10^0 \\ &-(1800+180+18) \\ &= 10^3+(2\cdot(9-\mathbf{1})+\mathbf{2})\cdot 10^2 \\ &+\ (2\cdot(9-\mathbf{2})+\mathbf{3})\cdot 10^1+2\cdot(9-\mathbf{3})\cdot 10^0 \\ &-(1800+180+18)\end{aligned}$$

But since the number 1998 (1800 plus 180 plus 18) can be put as 2000 minus 2 (*twice* $10^3 - 1$):

$$\begin{aligned}\mathbf{123} \cdot \mathbf{8} = 10^3 + (2 \cdot (9 - \mathbf{1}) + \mathbf{2}) \cdot 10^2 + (2 \cdot (9 - \mathbf{2}) + \mathbf{3}) \\ \cdot 10^1 + 2 \cdot (9 - \mathbf{3}) \cdot 1 - \underline{2} \cdot \left(10^3 - \underline{1}\right) \cdot 1 \\ = \left(\mathbf{1} - \underline{2}\right) \cdot 10^3 + (2 \cdot (9 - \mathbf{1}) + \mathbf{2}) \cdot 10^2 \\ + (2 \cdot (9 - \mathbf{2}) + \mathbf{3}) \cdot 10^1 + 2 \cdot \left(9 + \underline{1} - \mathbf{3}\right) \cdot 1 \\ = (\mathbf{1} - 2) \cdot 10^3 + (2 \cdot (9 - \mathbf{1}) + \mathbf{2}) \cdot 10^2 \\ + (2 \cdot (9 - \mathbf{2}) + \mathbf{3}) \cdot 10^1 + 2 \cdot (10 - \mathbf{3}) \cdot 10^0\end{aligned}$$

that is, when multiplying by 8, *each digit of the result can be obtained from the original number through addition, subtraction and duplication so that:*

- *The figure located furthest to the right of the result is* ***the double of the complement of 10*** *(subtraction of 10) of the units of the multiplicand.*
- *The remaining digits of the result (except the last) are calculated from right to left* ***ADDING to the double of the 9's complement del of the corresponding digit of the multiplicand*** *(the one with the same weight),* ***the one that precedes it*** *(the one on its*

*right) and the **carry**.*

- *The leftmost digit of the result is calculated by **subtracting 2** from the **leftmost digit** of the multiplicand and **adding the carry**.*

In our example:

$$\mathbf{123}\cdot\mathbf{8} = (\mathbf{1}-2)\cdot 10^3 + (2\cdot(9-\mathbf{1})+\mathbf{2})\cdot 10^2 + (2\cdot(9-\mathbf{2})+\mathbf{3})\cdot 10^1 + 2\cdot(10-\mathbf{3})\cdot 10^0 = (\mathbf{1}-1)\cdot 10^3 + (\mathbf{1}+8)\cdot 10^2 + (\mathbf{1}+7)\cdot 10^1 + 4\cdot 10^0 = \mathbf{0}\cdot 10^3 + \mathbf{9}\cdot 10^2 + \mathbf{8}\cdot 10^1 + \mathbf{4}\cdot 10^0 = \mathbf{0984}$$

where the carries have been transferred to the digit of the immediate higher power; let's explain it in words:

The $\mathbf{4}$ **of the units** comes from the $\boldsymbol{double\ of}$ $\mathbf{7}$ ($\mathbf{10}$ minus $\mathbf{3}$), that is, $\mathbf{14}$ ($\mathbf{4}$ units and $1\ carry$); the $\mathbf{8}$ **of the tens** is $\mathbf{18}$ ($\mathbf{8}$ tens and $1\ carry$) and is calculated by adding to $\mathbf{3}$ (the immediately preceding digit) twice the $9's\ complement\ of\ 2$ ($\mathbf{7}$ times $\mathbf{2}$) and $\mathbf{1}\ carry$; the $\mathbf{9}$ **of the hundreds** is $\mathbf{19}$ ($\mathbf{9}$ units and $1\ carry$) and is calculated by adding $\mathbf{2}$ (the immediately preceding digit of the multiplicand) to $twice\ the\ 9's\ complement\ of\ 1$ ($\mathbf{8}$

times **2**) plus **1** *carry*; and the **0 of the thousands** comes from subtracting **2** from **1** (the digit immediately before the multiplicand) and adding **1** *of carry*.

Another example is 891 (consider 0891 to make the algorithm easier) times 8, which results in **7128**:

The **8 of the units** is the ***double of*** **9** (**10** minus **1**), that is, **18** (**8** units and 1 *carry*); the **2 in the tens** comes from ***twice the*** **0** (the *9's complement* of 9) plus **1** (the immediately preceding digit) plus **1** *carry*; the **1 of the hundreds** comes from adding to **9** (the immediately preceding digit) ***the double of*** **1** (the *9's complement* of 8), namely **11** (**1** unit and 1 *carry*); finally, the **7 of the thousands** is obtained by **subtracting 2** from **8** (the leftmost digit of the multiplicand) and adding the immediately preceding carry (**6** plus **1** *of carry* is **7**).

Multiplication by 4

Multiplying by 4 will require some of the tricks used above, although it will be reduced to the simple calculation of the approximate half of a number and some additions and subtractions; we will explore the details shortly. The victim multiplicand is still 123 and the appearance of the traditional operation is as follows:

0	**1**	**2**	**3**
		×	**4**
0	**4**	**9**	**2**

The number **123**, is **100** (1 times 100) plus **20** (2 times 10) plus **3**; **multiplying it by 4 is** multiplying 100 by 4, 20 by 4 and 3 by 4; but since 5 minus 1 is 4 and half of 10 is exactly 5, we can say that:

$$\mathbf{123 \cdot 4} = (100 + 20 + 3) \cdot \left(\frac{10}{2} - 1\right)$$

$$= (100 + 20 + 3) \cdot \frac{10}{2} - (100 + 20 + 3) \cdot 1$$

$$= (1000 + 200 + 30) \cdot \frac{1}{2} - 100 - 20 - 3$$

$$= 1000 \cdot \frac{1}{2} - 100 + 200 \cdot \frac{1}{2} - 20 + 30 \cdot \frac{1}{2} - 3$$

Subtracting and adding 900, 90 and 9 (and grouping):

$$\mathbf{123 \cdot 4} = 1000 \cdot \frac{1}{2} - \mathbf{900} + \mathbf{900} - 100 + 200 \cdot \frac{1}{2} - \mathbf{90} + \mathbf{90}$$
$$- 20 + 30 \cdot \frac{1}{2} - \mathbf{9} + \mathbf{9} - 3$$
$$= 1000 \cdot \frac{1}{2} + \mathbf{900} - 100 + 200 \cdot \frac{1}{2} + \mathbf{90} - 20$$
$$+ 30 \cdot \frac{1}{2} + \mathbf{9} - 3 - \mathbf{900} - \mathbf{90} - \mathbf{9}$$
$$= \frac{1}{2} \cdot 10^3 + \left(\mathbf{9} - 1 + 2 \cdot \frac{1}{2}\right) \cdot 10^2$$
$$+ \left(\mathbf{9} - 2 + 3 \cdot \frac{1}{2}\right) \cdot 10^1 + (\mathbf{9} - 3) \cdot 10^0$$
$$- (\mathbf{900} + \mathbf{90} + \mathbf{9})$$

But since 999 (900 plus 90 plus 9) can be put as 1000 (10 cubed) minus 1:

$$\mathbf{123 \cdot 4} = \frac{1}{2} \cdot 10^3 + \left(9 - 1 + \frac{2}{2}\right) \cdot 10^2 + \left(9 - 2 + \frac{3}{2}\right) \cdot 10^1$$
$$+ (9 - 3) \cdot 10^0 - (\mathbf{10^3} - \mathbf{1})$$
$$= \left(\frac{1}{2} - \mathbf{1}\right) \cdot 10^3 + \left(9 - 1 + \frac{2}{2}\right) \cdot 10^2$$
$$+ \left(9 - 2 + \frac{3}{2}\right) \cdot 10^1 + (9 - 3 + \mathbf{1}) \cdot 10^0$$
$$= \left(\frac{\mathbf{1}}{2} - 1\right) \cdot 10^3 + \left(9 - \mathbf{1} + \frac{2}{2}\right) \cdot 10^2$$
$$+ \left(9 - \mathbf{2} + \frac{\mathbf{3}}{2}\right) \cdot 10^1 + (10 - \mathbf{3}) \cdot 10^0$$

If the digit that is divided by 2 is odd, it is necessary to add 0.5 (the fractional part that is generated), which is equivalent to adding 5 to the immediate term with the least weight. In this case 1 and 3 are odd, so their corresponding fractions are replaced by the quotient when dividing by 2 y and the fractional part (5) is transferred to the adjacent term with the least weight, so the previous expression becomes:

$$\begin{aligned}
\mathbf{123}\cdot\mathbf{4} &= \left(\frac{\mathbf{1}}{2}-1\right)\cdot 10^3+\left((9-\mathbf{1})+\frac{\mathbf{2}}{2}\right)\cdot 10^2 \\
&\quad+\left((9-\mathbf{2})+\frac{\mathbf{3}}{2}\right)\cdot 10^1+(10-\mathbf{3}) \\
&= (\mathbf{0}-1)\cdot 10^3+\left((9-\mathbf{1})+1+\mathbf{5}\right)\cdot 10^2 \\
&\quad+\left((9-\mathbf{2})+\mathbf{1}\right)\cdot 10^1+\left((10-\mathbf{3})+\mathbf{5}\right)\cdot 10^0 \\
&= (-1+\mathbf{1})\cdot 10^3+\mathbf{4}\cdot 10^2+(8+\mathbf{1})\cdot 10^1+\mathbf{2} \\
&= \mathbf{0}\cdot 10^3+\mathbf{4}\cdot 10^2+\mathbf{9}\cdot 10^1+\mathbf{2}\cdot 10^0=\mathbf{0492}
\end{aligned}$$

where the carries have been transferred from right to left to the digit of the next higher power; when we multiply by 4, *each digit of the result can be deducted from the original number as follows:*

- *The rightmost figure of the result is obtained by calculating the* ***10's complement*** *(subtraction of 10)* ***from the units of the multiplicand*** *and adding* ***5*** *(only* ***if*** *said* ***units are*** *an* ***odd*** *number).*
- *The remaining digits of the result (except the last) are calculated from right to left* ***by adding*** *to the* ***9's complement of the corresponding digit of the multiplicand*** *(the one with the same weight), the* ***TRUNCATED HALF of the preceding digit*** *(the one on its right),* ***5*** *(only when the digit of the multiplicand is odd) and the* ***carry*** *(if applicable).*
- *The leftmost digit of the result is calculated* ***by SUBTRACTING 1*** *from the* ***TRUNCATED HALF of the leftmost digit*** *of the multiplicand and* ***adding the carry*** *(if necessary).*

Let us explain in words, with the help of the example **123 · 4**, the relationship between each digit of the result of the operation (**0492**) and each digit of the multiplicand:

The **2 in the ones** comes from **12** (**2** units and

1 *carry*) and is the sum of **5** (the 3 of the multiplicand is odd) and the 10′*s complement* of 3 (**10** minus **3**); the **9 of the tens** is calculated by adding **7** (9′*s complement* of 2) plus **0** (since the 2 in the multiplicand is even, 5 is not added) plus **1** (the truncated quotient of dividing 3 by 2) plus **1** *carry*); the **4 of the hundreds** is the sum of **5** (the 1 of the multiplicand is odd) plus **1** (the truncated quotient of dividing the digit 2 of the multiplicand by 2) plus **8** (9′*s complement* of 1), resulting in **14** (**4** units and 1 *carry*); finally, the **0 in the thousands** is: **0** (the truncated half of 1) plus **1** *carry* minus **1**.

Another example is **891 · 4** whose result is **3564**:

The **4 of the units** comes from **14** (**4** units and 1 *carry*) —**add 5** (because the 1 in the multiplicand is odd) to the 10′*s complement* of 1 (**10** minus **1**)—; the **6 of the tens** is: **0** (*the* 9′*s complement of* 9) plus **5** (since the 9 of the multiplicand is odd, 5 is added) plus **0** (truncated quotient of dividing 1 by 2) plus **1** *carry*; the **5 of the hundreds** is: **0** (the **8** of the multiplicand is even) plus **1** (**9** minus **8**) plus **4** (the truncated quotient of dividing **9** by **2**); finally, the **3 in the thousands** is **4** (the truncated half of 8) minus **1**.

Multiplication by 3

Having a multiplier of 3 is similar to multiplying by 4; the algorithm is almost identical; it only requires additionally doubling some number before adding it. As multiplicand we will use, as always, 123. The traditional way of operating looks like this:

0	**1**	**2**	**3**
		×	**3**
0	**3**	**6**	**9**

The number **123**, is **100** (1 times 100) plus **20** (2 times 10) plus **3**; **multiplying it by 3 is** multiplying 100 by 3, 20 by 3, and 3 by 3 (and adding); but since 5 minus 2 is 3 and half of 10 is exactly 5, we can say that:

$$\begin{aligned}\mathbf{123\cdot 3} &= (100+20+3)\cdot\left(\frac{10}{2}-2\right)\\ &= (100+20+3)\cdot\frac{10}{2}-(100+20+3)\cdot 2\\ &= (1000+200+30)\cdot\frac{1}{2}-200-40-6\\ &= 1000\cdot\frac{1}{2}-200+200\cdot\frac{1}{2}-40+30\cdot\frac{1}{2}-6\end{aligned}$$

Subtracting and adding 1800, 180, and 18 (twice 900, 90 and 9, respectively) and grouping:

$$\begin{aligned}\mathbf{123\cdot 3} &= 1000\cdot\frac{1}{2}-\mathbf{1800}+\mathbf{1800}-200+200\cdot\frac{1}{2}-\mathbf{180}\\ &\quad+\mathbf{180}-40+30\cdot\frac{1}{2}-\mathbf{18}+\mathbf{18}-6\\ &= 1000\cdot\frac{1}{2}+\mathbf{1800}-200+200\cdot\frac{1}{2}+\mathbf{180}\\ &\quad-40+30\cdot\frac{1}{2}+\mathbf{18}-6-\mathbf{1800}-\mathbf{180}-\mathbf{18}\\ &= \frac{1}{2}\cdot 10^3+\left(\mathbf{18}-2+2\cdot\frac{1}{2}\right)\cdot 10^2\\ &\quad+\left(\mathbf{18}-4+3\cdot\frac{1}{2}\right)\cdot 10^1+(\mathbf{18}-6)\cdot 10^0\\ &\quad-(\mathbf{1800}+\mathbf{180}+\mathbf{18})\end{aligned}$$

Since the number 1998 (twice the sum of 900 plus 90 plus 9) equals 2000 (twice 10 *cubed*) minus 2:

$$\mathbf{123 \cdot 3} = \frac{1}{2} \cdot 10^3 + \left(18 - 2 + \frac{2}{2}\right) \cdot 10^2 + \left(18 - 4 + \frac{3}{2}\right) \cdot 10^1$$
$$+ (18 - 6) \cdot 10^0 - (\mathbf{2 \cdot 10^3 - 2}) \cdot 10^0$$
$$= \left(\frac{1}{2} - \mathbf{2}\right) \cdot 10^3 + \left(18 - 2 + \frac{2}{2}\right) \cdot 10^2$$
$$+ \left(18 - 4 + \frac{3}{2}\right) \cdot 10^1 + (18 - 6 + \mathbf{2}) \cdot 10^0$$
$$= \left(\frac{1}{2} - \mathbf{2}\right) \cdot 10^3 + \left(2 \cdot 9 - 2 + \frac{2}{2}\right) \cdot 10^2$$
$$+ \left(2 \cdot 9 - 4 + \frac{3}{2}\right) \cdot 10^1 + (2 \cdot 9 - 4) \cdot 10^0$$
$$= \left(\frac{\mathbf{1}}{2} - 2\right) \cdot 10^3 + \left(2 \cdot (9 - \mathbf{1}) + \frac{\mathbf{2}}{2}\right) \cdot 10^2$$
$$+ \left(2 \cdot (9 - \mathbf{2}) + \frac{\mathbf{3}}{2}\right) \cdot 10^1 + 2 \cdot (10 - \mathbf{3}) \cdot 10^0$$

If the digit that is divided by 2 is odd, it is necessary to add 0.5 (the fractional part that is generated), which is equivalent to adding 5 to the immediate term with the least weight. In this case 1 and 3 are odd, so their corresponding fractions are replaced by the quotient when dividing by 2 and the fractional part (5) is transferred to the adjacent term with the least weight, so the previous expression results in:

$$\mathbf{123} \cdot \mathbf{3} = \left(\frac{\mathbf{1}}{2} - 2\right) \cdot 10^3 + \left(2 \cdot (9 - \mathbf{1}) + \frac{\mathbf{2}}{2}\right) \cdot 10^2$$
$$+ \left(2 \cdot (9 - \mathbf{2}) + \frac{\mathbf{3}}{2}\right) \cdot 10^1 + 2 \cdot (10 - \mathbf{3}) \cdot 10^0$$
$$= (\mathbf{0} - 2) \cdot 10^3 + (2 \cdot (9 - 1) + 1 + \mathbf{5}) \cdot 10^2$$
$$+ \ (2 \cdot (9 - 2) + \mathbf{1}) \cdot 10^1 + (2 \cdot (10 - 3) + \mathbf{5})$$
$$= (-2 + \mathbf{2}) \cdot 10^3 + (2 + \mathbf{1}) \cdot 10^2 + (5 + \mathbf{1}) \cdot 10$$
$$+ 9 = \mathbf{0369}$$

where the carries have been moving to the digit of the immediate higher power; that is, it is true that when multiplying by 3, *each digit of the result can be obtained from the original number as follows:*

- *The rightmost figure of the result is obtained by calculating the* ***DOUBLE of the 10's complement*** *(twice the subtraction 10)* ***from the units of the multiplicand*** *and adding* ***5*** *(only* ***if*** *said units* ***are*** *an* ***odd*** *number).*
- *The remaining figures of the result (except the last) are calculated from right to left* ***by adding*** *to the* ***DOUBLE of the 9's complement of the digit of the same position*** *of the multiplicand, the* ***TRUNCATED***

HALF of the one that precedes it (the one on its right) plus **5** (only when the digit of the multiplicand is odd) plus the **carry** (if applicable).*

- *The leftmost digit of the result is calculated* ***by SUBTRACTING 2*** *from the* ***TRUNCATED HALF of the leftmost digit*** *of the multiplicand and* ***adding the carry*** *(if necessary).*

The example $\mathbf{123 \cdot 3}$ helps to see the relationship between each digit of the result of the operation ($\mathbf{0369}$) and each digit of the multiplicand:

The $\mathbf{9}$ **of the units** comes from $\mathbf{19}$ ($\mathbf{9}$ units and $1\,carr$) value obtained by adding $\mathbf{5}$ (since the 3 of the multiplicand is odd) to **twice the** $\mathbf{10's\ complement}$ of 3 (10 minus 3 is 7, times 2 is $\mathbf{14}$); the sum "$\mathbf{14}$ (**twice the** $\mathbf{9's\ complement}$ of 2) plus $\mathbf{0}$ (because the 2 in the multiplicand is even) plus $\mathbf{1}$ (the truncated quotient of dividing 3 by 2) plus $\mathbf{1}\ carry$" totals $\mathbf{16}$ ($\mathbf{6}$ units and $1\,carry$) determining the $\mathbf{6}$ **of the tens**; the sum "$\mathbf{16}$ (**twice the** $\mathbf{9's\ complement}$ of 1) plus $\mathbf{5}$ (the 1 in the multiplicand is odd) plus $\mathbf{1}$ (the truncated quotient of dividing 3 by 2) plus **1** *carry*, totals **23** (**3** units and 2 *carry*) defining the **3 of the hundreds**; finally, the **0 of the thousands** is **0** (the truncated half of 1) **plus 2** (from the previous carry) **minus 2**.

Another example is 891 · 3 whose result is **2673**:

The **3 of the units** comes from **23** (**3** units and 2 *carry*), value obtained by adding **5** (since the 1 of the multiplicand is odd) to **twice the 10's *complement*** of 1 (10 minus 1 is 9, times 2 is **18**); the sum "**0** (twice the 9's *complement* of 9) plus **5** (since the 9 in the multiplicand is odd, 5 is added) plus **0** (the truncated quotient of dividing 1 by 2) plus **2** *carry*" gives the **7 of the tens**; adding "**0** (the 8 of the multiplicand is even) plus **2** (9 minus 8, times 2) plus **4** (el the quotient when dividing 9 by 2)" we determine the **6 of the hundreds**; and finally, the **2 of the thousands** is **4** (the truncated half of 8) **minus 2**.

Multiplication by 2

Having 2 as a multiplier is very simple; only requires twice each digit of the multiplicand; the previous digit is not required, just the sum of the generated carry.

The multiplicand we are going to use is 689. The traditional way of operating looks like this:

0	**6**	**8**	**9**
		×	**2**
1	**3**	**7**	**8**

The number **689** is **600** (6 times 100 or $6 \cdot 10^2$) plus **80** (8 times 10 or $8 \cdot 10^1$) plus **9** (9 times 1 or $9 \cdot 10^0$); **multiplying it by 2 is** multiplying **600 by 2**, **80 by 2** and **9 by 2** (and adding):

$$\begin{aligned} \mathbf{689} \cdot \mathbf{2} &= (600 + 80 + 9) \cdot 2 \\ &= (\mathbf{6} \cdot 10^2 + \mathbf{8} \cdot 10^1 + \mathbf{9} \cdot 10^0) \cdot 2 \\ &= (2 \cdot \mathbf{6} \cdot 10^2 + 2 \cdot \mathbf{8} \cdot 10^1 + 2 \cdot \mathbf{9} \cdot 10^0) \end{aligned}$$

when carry occurs, it must be transmitted to the

immediately higher weighting digit:

$$\begin{aligned}\mathbf{689} \cdot \mathbf{2} = (0 \cdot 10^3 + 2 \cdot 6 \cdot 10^2 + 2 \cdot 8 \cdot 10^1 + 2 \cdot 9 \cdot 10^0) \\ = (0 + \mathbf{1}) \cdot 10^3 + (2 + \mathbf{1}) \cdot 10^2 + (6 + \mathbf{1}) \cdot 10^1 \\ + 8 \cdot 10^0 = \mathbf{1} \cdot 10^3 + \mathbf{3} \cdot 10^2 + \mathbf{7} \cdot 10^1 + \mathbf{8} \cdot 10^0 \\ = \mathbf{1378}\end{aligned}$$

from which the following rule can be inferred:

When we multiply by 2, *each digit of the result can be obtained from the original number (we add a 0 on the left to be able to generalize) simply by calculating the* ***double of the digit of the same position*** *of the multiplicand and adding the immediately previous carry (if exists).*

Next, all the operations of $0\mathbf{689} \cdot \mathbf{2}$, whose result is $\mathbf{1378}$, explained in detail:

The **8 of the units** comes from twice 9, that is, **18** (**8** units and 1 *carry*); the **7 of the tens** is the sum of **16** (twice 8) and **1** *carry*, which equals **17** (**7** units and 1 *carry*); finally, **12** (twice 6) plus **1** *carry* is **13** (that is, **3** units that constitute **the tens of the result** and 1 *carry* that becomes the **1** of the thousands when added to twice 0).

Multiplication by 1

Multiplying by 1 consists of repeating the multiplicand and offering it as a result.

Multiplication by 0

Multiplying by 0 is even easier: it's always 0.

Rules Summary

We have seen that when using each of the numbers as a multiplier there are several basic *tricks* that we can use in combination to deduce each digit of the result from the one with the same position of the multiplicand, the carry and the immediately previous one (the neighbor to its right), namely:

- Add the immediately preceding digit.
- Multiply a digit by 2.
- Add half (truncated) of a number.
- Add 5 if the considered digit is odd.
- Add the immediately preceding carry.

- Subtract the considered digit from 9 or 10.

Multiplication by 12: add to the DOUBLE of the current digit the one to its right and the carry. $P.ej. 0\mathbf{127} \cdot 12 = \mathbf{1524}$:

$$\begin{array}{lll r} 2 \cdot \mathbf{7} + 0 & = 4\ (carry\ 1) & & 4 \\ 2 \cdot \mathbf{2} + \mathbf{7} + (1) & = 2\ (carry\ 1) & & 24 \\ 2 \cdot \mathbf{1} + \mathbf{2} + (1) & = 5\ (carry\ 0) & & 524 \\ 2 \cdot 0 + \mathbf{1} + (0) & = 1\ (carry\ 0) & & \mathbf{1524} \end{array}$$

Multiplication by 11: add to the current digit the one to its right and the carry. $P.ej. 0\mathbf{157} \cdot 11 = \mathbf{1727}$:

$$\begin{array}{lll r} \mathbf{7} + 0 & = 7\ (carry\ 0) & & 7 \\ \mathbf{5} + \mathbf{7} + (0) & = 2\ (carry\ 1) & & 27 \\ \mathbf{1} + \mathbf{5} + (1) & = 7\ (carry\ 0) & & 727 \\ 0 + \mathbf{1} + (0) & = 1\ (carry\ 0) & & \mathbf{1727} \end{array}$$

Multiplication by 10: no calculations required; the result of the operation is the multiplicand with one more zero to its right.

Multiplication by 9: the **least significant digit** is obtained by subtracting the units of the multiplicand from 10; the **intervening digits**, subtracting from 9 the current digit of the multiplicand and adding the one to its right and the carry; and the **most significant digit**, subtracting 1 from the most significant digit of the multiplicand and adding the carry. $P.ej. \mathbf{142} \cdot 9 = \mathbf{1278}$:

$$\begin{array}{rll} (10-\mathbf{2}) & = 8\ (carry\ 0) & 8 \\ (9-\mathbf{4})+\mathbf{2}+(0) & = 7\ (carry\ 0) & 78 \\ (9-\mathbf{1})+\mathbf{4}+(0) & = 2\ (carry\ 1) & 278 \\ (\mathbf{1}-1)+(1) & = 1\ (carry\ 0) & \mathbf{1278} \end{array}$$

Multiplication by $\mathbf{8}$: for the **least significant digit**, the units of the multiplicand are subtracted from 10 and the result is multiplied by 2; for **middle digits**, the current digit of the multiplicand is subtracted from 9 (and is multiplied by 2), adding the digit to its right, and the carry; and for the **most significant digit,** the carry is added to the most significant digit minus 2 of the multiplicand. $P.ej.\mathbf{5238}\cdot 8=\mathbf{41904}$:

$$\begin{array}{rll} (10-\mathbf{8})\cdot 2 & = 4\ (carry\ 0) & 4 \\ (9-\mathbf{3})\cdot 2+\mathbf{8}+(0) & = 0\ (carry\ 2) & 04 \\ (9-\mathbf{2})\cdot 2+\mathbf{3}+(2) & = 9\ (carry\ 1) & 904 \\ (9-\mathbf{5})\cdot 2+\mathbf{2}+(1) & = 1\ (carry\ 1) & 1904 \\ (\mathbf{5}-2)+(1) & = 4\ (carry\ 0) & \mathbf{41904} \end{array}$$

Multiplication by $\mathbf{7}$: must be added to the DOUBLE of the current digit 5 (only if said digit is odd), the truncated half of the one to its right and the carry. $P.ej.0\mathbf{4163}\cdot 7=\mathbf{29141}$:

$$\begin{array}{rll} 2\cdot\mathbf{3}+5+[\mathbf{0}/2] & = 1\ (carry\ 1) & 1 \\ 2\cdot\mathbf{6}+0+[\mathbf{3}/2]+(1) & = 4\ (carry\ 1) & 41 \\ 2\cdot\mathbf{1}+5+[\mathbf{6}/2]+(1) & = 1\ (carry\ 1) & 141 \\ 2\cdot\mathbf{4}+0+[\mathbf{1}/2]+(1) & = 9\ (carry\ 0) & 9141 \\ 2\cdot 0+0+[\mathbf{4}/2]+(0) & = 2\ (carry\ 0) & \mathbf{29141} \end{array}$$

Multiplication by $\mathbf{6}$: you have to add to the current digit 5 (only if it is odd), half of the one to its right (truncated) and the carry. $P.ej. 0\mathbf{5273} \cdot 6 = \mathbf{31638}$:

$$\begin{array}{rll} \mathbf{3} + 5 + [\mathbf{0}/2] & = 8\ (carry\ 0) & 8 \\ \mathbf{7} + 5 + [\mathbf{3}/2] + (0) & = 3\ (carry\ 1) & 38 \\ \mathbf{2} + 0 + [\mathbf{7}/2] + (1) & = 6\ (carry\ 0) & 638 \\ \mathbf{5} + 5 + [\mathbf{2}/2] + (0) & = 1\ (carry\ 1) & 1638 \\ 0 + 0 + [\mathbf{5}/2] + (1) & = 3\ (carry\ 0) & \mathbf{31638} \end{array}$$

Multiplication by $\mathbf{5}$: it is enough to add 5 (only if the current digit is odd) to the half (truncated) of the one to its right. $P.ej. 0\mathbf{367} \cdot 5 = \mathbf{1835}$:

$$\begin{array}{rll} 5 + [\mathbf{0}/2] & = 5\ (current\ digit\ 7 — odd) & 5 \\ 0 + [\mathbf{7}/2] & = 3\ (current\ digit\ 6 — even) & 35 \\ 5 + [\mathbf{6}/2] & = 8\ (current\ digit\ 3 — odd) & 835 \\ 0 + [\mathbf{3}/2] & = 1\ (current\ digit\ 0 — even) & \mathbf{1835} \end{array}$$

Multiplication by $\mathbf{4}$: the **least significant digit** is calculated by subtracting the units of the multiplicand from 10 and adding 5 (only when they are an odd number); the **intermediate digits**, subtracting from 9 the current digit of the multiplicand, and adding 5 (only if it is odd), half (truncated) of the one to its right and the carry; finally, the **most significant digit** is determined by subtracting 1 from the leftmost (truncated) half of the multiplicand and adding the carry. $P.ej. \mathbf{921} \cdot 4 = \mathbf{3684}$:

$$\begin{array}{rlr} (10-\mathbf{1})+5 & =4\ (carry\ 1) & 4 \\ (9-\mathbf{2})+0+[\mathbf{1}/2]+(1) & =8\ (carry\ 0) & 84 \\ (9-\mathbf{9})+5+[\mathbf{2}/2]+(0) & =6\ (carry\ 0) & 684 \\ [\mathbf{9}/2]-1+(0) & =3\ (carry\ 0) & \mathbf{3684} \end{array}$$

Multiplication by 3: the **least significant digit** is calculated by doubling the subtraction of 10 from the units of the multiplicand and adding 5 (if these are an odd number); the **middle digits**, by doubling the subtraction of 9 from the current digit of the multiplicand, and adding 5 (if it is an odd number), half (truncated) of the one to its right and the carry; finally, the **most significant digit** is calculated by subtracting 2 from the (truncated) half of the most significant digit of the multiplicand and adding the carry. $P.ej.\mathbf{721}\cdot 3=\mathbf{2163}$:

$$\begin{array}{rlr} (10-\mathbf{1})\cdot 2+5 & =3\ (carry\ 2) & 3 \\ (9-\mathbf{2})\cdot 2+0+[\mathbf{1}/2]+(2) & =6\ (carry\ 1) & 63 \\ (9-\mathbf{7})\cdot 2+5+[\mathbf{2}/2]+(1) & =1\ (carry\ 1) & 163 \\ [\mathbf{7}/2]-2+(1) & =2\ (carry\ 0) & \mathbf{2163} \end{array}$$

Multiplication by 2: add the current digit twice and add the corresponding carry. $P.ej.0\mathbf{597}\cdot 2=\mathbf{1194}$:

$$\begin{array}{llr} \mathbf{7}+\mathbf{7} & =4\ (carry\ 1) & 4 \\ \mathbf{9}+\mathbf{9}+(1) & =9\ (carry\ 1) & 94 \\ \mathbf{5}+\mathbf{5}+(1) & =1\ (carry\ 1) & 194 \\ 0+0+(1) & =1\ (carry\ 0) & \mathbf{1194} \end{array}$$

Multiplication by 1: no calculation required; the result of the operation is identical to the multiplicand.

Multiplication by 0: multiplying a number by 0 we always get 0 as a result.

Chapter 2

Fast multiplication

The objective of this chapter is to be able to multiply multi-digit numbers directly, without the need to write the entire process of traditional multiplication.

Two-digit multiplier

First of all, we are going to explore the operations to be carried out in a multiplication with a *multiplicand and a two-digit multiplier.*

- The **units** *of the result* come from the product of the units of each of the operands.
- The **tens** *of the result* are the sum of the products of the units of each of the operands by the tens of the other (and the immediately preceding

carry).

- The **hundreds** *of the result* comes from the product of the tens of the operands plus the immediately preceding carry.

Traditionally, this is expressed like this:

0	0	**4**	**3**
	×	**2**	**1**
		$(1 \times 4) = \mathbf{4}$	$(1 \times 3) = \mathbf{3}$
	$(2 \times 4) = \mathbf{8}$	$(2 \times 3) = \mathbf{6}$	
	$\mathbf{8} + 1(carry) = \mathbf{9}$	$(\mathbf{6} + \mathbf{4}\ is\ \mathbf{10}) = \mathbf{0}$	**3**

If we put the multiplicand in line after the multiplier and *number from 1 to 4 and from right to left the positions,* that is, (43 21) the algorithm that solves the product translates into these steps:

- To obtain the **units** *of the result*, it is enough to multiply positions 1 and 3.
- To find the **tens** *of the result*, we add to the immediately previous carry (if it exists) the

products of the numbers located in the extreme and middle positions ($1\ times\ 4$ and $2\ times\ 3$).

- The **hundreds** and **thousands** are determined by adding the carry to the product of the digits in positions 2 and 4.

For example, when multiplying $89\ by\ 13$ (whose result is **1157**) the following steps do the job:

- The **7 of the ones** comes from $9\ times\ 3$, which is **27** (**7** units and $2\ carry$).
- For the **5 of the tens** the digits of extreme and middle positions are multiplied ($8\ times\ 3$ is **24** and $1\ times\ 9$ is **9**); both results plus **2** *of carry* make a total of **35** (**5** units and $3\ carry$).
- The **11 of hundreds and thousands** is calculated by adding to the product of $1\ times\ 8$ (which is **8**), **3** *carry*.

Now let's consider the case where the *multiplicand has more than two digits*. The procedure to follow is analogous to the previous one. The *rightmost digit* of the result is calculated identically; in the *central digits* we have to use the same algorithm, but always using 4 numbers, namely, the two of the multipliers and two of the multiplicand (the one with the same position as the digit of the result in the process of calculation and the immediately previous one). If we add *two zeros* to the left of the multiplicand, the same method can be used to obtain the *leftmost digit* of the result. Let's clear this all up a bit. Traditionally the operation looks something like this:

0	0	**5**	**4**	**3**
		×	**2**	**1**
		5	4	3
1	0	8	6	
1	**1**	**4**	**0**	**3**

The *two zeros* added to the left allow us to remember when to end the multiplication process:

The **3 of the units** comes from multiplying the 1 of the units of the multiplier by the units of the multiplicand (1 *times* 3 is **3**); to calculate the **0 of the tens** we proceed by selecting the four digits that must come into play (43 21) and we make the sum of the products of the extreme elements (1 *times* 4 is **4**) and middle elements (2 *times* 3 are **6**), that is, **10** (**0** units and 1 *carry*); for the **4 of the hundreds** the four digits to consider are (54 21); the product of extreme elements is **5** (1 *times* 5) and that of the middle elements is **8** (2 *times* 4) which, added to **1** *carry* make **14** (**4** units and 1 *carry*); the **1 of the thousands** is calculated from the corresponding four digits (05 21); the sum of the extreme (1 *times* 0 is **0**) and middle products (2 *times* 5 are **10**) plus **1** *carry* is **11** (**1** and 1 *carry*); the leftmost 1 of the result is trivial to obtain because this time the four digits are (00 21) and since 0 times any number is 0, the sum of the products of the extreme and middle elements is 0 and therefore so we only have to add **1** *carry*.

The algorithm used in the previous paragraph to multiply by a two-digit number is equally valid when the multiplicand has more than three digits. The number of zeros to add to the left is still two and the calculation is identical.

When multiplying and/or multiplier have zeros to the right, the multiplication can be simplified by transferring them to the result and ignoring said zeros when operating; e.g., 4500 times 20 is 90 000 and is equivalent to the operation: 45 times 2 and multiply the result (90) by 1000 (i.e., add three zeros).

Three-digit multiplier

As will be seen shortly, the procedure to follow is similar to operating with a two-digit multiplier with small variations; the zeros to add to the left of the multiplicand to facilitate the algorithm are three instead of two. Let's take as an example, **654** times **321**.

The **rightmost digit** of the result is calculated by multiplying the rightmost digit of the multiplicand by the analog of the multiplier (1 *times* 4 is **4**).

The **tens** of the result are calculated in the same

way as it was done with a two-digit multiplier, that is, we choose the 2 least significant digits of the multiplicand and the two least significant digits of the multiplier (54 21), adding the products of the extremes (5 *por* 1) and middles (4 *por* 2) values; this gives **13** as a result (**3** units and 1 *carry*).

The **remaining digits** of the result require sets of six numbers chosen in such a way that, being the three of the multiplier always fixed, the three of the multiplicand vary as they are taken from right to left with a shift of one digit *(the initial shift is zero).* Putting the six selected digits in line, each digit of the result will be the sum of the three products obtained whose factors are calculated in mirror (from outside to inside, each of the multiplier times the one among those chosen from the multiplicand whose mirror position corresponds). Let's clear all this up: for the **hundreds** of result the group of six numbers is (654 321); the mirror positions of 1, 2 and 3 are respectively 6, 5 and 4; the addends are **6** (6 *times* 1), **10** (5 *times* 2) and **12**

(4 *times* 3) whose sum is **28**, which *after adding the carry gives* **29** (**9** units and 2 *carry*). For the **units of thousands** of the result, the group of six numbers is (065 321); the mirror positions of 1, 2 and 3 are respectively 0, 6 and 5; the addends are therefore **0** (0 *times* 1), **12** (6 *times* 2) and **15** (5 *times* 3) whose sum is **27**, which plus 2 *carry* is **29** (**9** units and 2 *carry*). For the **tens of thousands** of the result the six numbers are (006 321); the mirror positions of 1, 2 and 3 are respectively 0, 0 and 6; the addends are therefore **0** (0 *times* 1), **0** (0 *times* 2) y **18** (6 *times* 3) whose sum is **18**, which plus 2 *carry* is **20** (**0** units and 2 *carry*). Finally, for the **hundreds of thousands** of the result, the group of six numbers is (000 321); the mirror positions of 1, 2 and 3 are respectively 0, 0 and 0; the addends are therefore all **0** as well as their sum, whichever plus 2 *carry* is **2** (**2** units and 0 *carry*).

Collecting information, we see that the algorithm used to multiply 654 by 321 gives the traditionally expected result (209 934):

0	0	0	**6**	**5**	**4**
		×	**3**	**2**	**1**
			6	5	4
	1	3	0	8	
1	9	6	2		
2	**0**	**9**	**9**	**3**	**4**

Multiplier of 4 or more digits

The method to follow can be deduced from the procedure used above when we operate with a three-digit multiplier. As an example, we are going to multiply **98 765** (the multiplicand) by **4321** (the multiplier); the result should give us **426 763 565**.

To give generality to the algorithm, we consider the multiplicand preceded by *as many zeros as digits there are in the multiplier;* in this case 4321 has 4 digits which converts the multiplicand into **000 098 765**. Each digit of the result comes from the sum of a certain number of factors: For the **units** of the result, the rightmost digit of the multiplier and the analog of the multiplicand are required (5 *times* 1 is **5**). The two rightmost digits of the multiplicand and multiplier (65 21) produce the 2 factors of each addend (6 *times* 1 of the extremes and 5 *times* 2 of the middles) that constitute the **tens** of the result (6 plus 10 are **16**, that is, **6** units and 1 *carry*). For the **hundreds**, three addends generated in the same way as before are required; the digits to be considered are the three least significant of the multiplier (765) and of the multiplicand (321) which, placed in line (765 321) allow us to know the two factors of each addend without more than taking the two extreme values towards the interior (7 *times* 1, 6 *times* 2 and 5 *times* 3), which added to the carry (7 plus 12 plus 15 plus 1 *carry*) make a total of **35** (**5** units and 3 *carry*). The **units of thousand** follow the same procedure; the digits to consider are this time (8765 4321); the two factors of each of the four addends are chosen as before (8 *times* 1, 7 *times* 2, 6 *times* 3 y 5 *times* 4), which added to the immediately preceding carry (8 plus 14 plus 18 plus 20 plus 3 *carry*) make a total of **63** (**3** units and 6 *carry*). From here, and because there are no more digits in the multiplier, all its digits will be taken and as many of the multiplicand as there are figures in the multiplier, selected with a shift to the left until the zeros that we had added to the multiplicand at the beginning of the entire process have been used: (9876 4321), (0987 4321), (0098 4321), (0009 4321) y (0000 4321). More in detail:

9876 4321 generates the 2 factors of each of the four addends (9 *times* 1, 8 *times* 2, 7 *times* 3 and 6 *times* 4), which together with the previous carry (9 plus 16 plus 21 plus 24 plus 6 *carry*) are **76** (**6** units and 7 *carry*).

0987 4321 generates the 2 factors of each of the four addends (0 *times* 1, 9 *times* 2, 8 *times* 3 and 7 *times* 4), which together with the previous carry (0 plus 18 plus 24 plus 28 plus 7 *carry*) are **77** (**7** units and 7 *carry*).

0098 4321 generates the 2 factors of each of the four addends (0 *times* 1, 0 *times* 2, 9 *times* 3 and 8 *times* 4), which together with the previous carry (0 plus

0 plus 27 plus 32 plus 7 *carry*) are **66** (**6** units and 6 *carry*).

0009 4321 generates the 2 factors of each of the four addends (0 *times* 1, 0 *times* 2, 0 *times* 3 and 9 *times* 4), which together with the previous carry (0 plus 0 plus 0 plus 36 plus 6 *carry*) are **42** (**2** units and 4 *carry*).

0000 4321 generates the two factors of each of the four addends (0 *times* 1, 0 *times* 2, 0 *times* 3 and 0 *times* 4), which together with the previous carry (0 plus 0 plus 0 plus 0 plus 4 *carry*) are **4**.

This algorithm can be generalized; it is enough to add to the left of the multiplicand as many zeros as figures there are in the multiplier to determine when to end the process and once, we run out of digits of the multiplier, continue choosing as many figures of the multiplicand as there are in the multiplier with a shift to the left, as already was done above.

Chapter 3

Two-finger method

This method requires clarifying some concepts and setting some basic rules:

- The word digit refers to a single digit number.
- The maximum product of two digits never occupies more than two digits (9 *times* 9 is 81).
- If a digit needs to be treated as a two-digit number, it is considered preceded by a zero, that is, it will have a zero as tens and the digit itself as units.
- In two-digit numbers, the units will be identified by the letter ***U***; for the tens, will be used the letter ***D***. When multiplying two digits, sometimes only the tens will be used and other times only the

units. To indicate it, a pattern formed by the letters ***U*** (take *only the units*), ***D*** (take *only the tens*) and ***o*** (ignore digit) will be used. This pattern will be applied exclusively to the multiplicand and will determine (by analogy of location) the factors that generate each addend of the operation in progress. For example, if the multiplicand is **182** and the multiplier is **4**:

- The pattern ***U*** indicates to take the *units* of 2 *times* 4 (that is, 8).
- The ***UD*** pattern tells us to add the *tens* of 2 *times* 4 (the **0** of 08) to the *units* of 8 *times* 4 (the **2** of 32), that is, 0 plus 2 is **2**.
- The ***UDo*** pattern tells us to ignore the units digit of the multiplicand and add the *tens* of 8 *times* 4 (the **3** of 32) to the *units* of 1 *times* 4 (the **4** of 04), that is, **3** plus **4** is **7**.

Once one has become sufficiently familiar with these concepts, the algorithm used by the two-finger method becomes simple and practical.

To clarify ideas, we are going to see algebraically why the method of units and tens that we have called here *of two-finger* works. The cba number consists of three figures and is (always in base $base$ 10):

$$\boldsymbol{cba} = c \cdot 10^2 + b \cdot 10^1 + a \cdot 10^0 = \boldsymbol{c \cdot 100 + b \cdot 10 + a}$$

Multiplying a, b or c by a single-digit number n produces a number of at most two digits (the maximum value would be 9 times 9, which is 81), namely:

$$\begin{aligned}
\boldsymbol{a} \cdot n &= \boldsymbol{D}_a \cdot 10^1 + \boldsymbol{U}_a \cdot 10^0 = \boldsymbol{D}_a \cdot 10 + \boldsymbol{U}_a \\
\boldsymbol{b} \cdot n &= \boldsymbol{D}_b \cdot 10^1 + \boldsymbol{U}_b \cdot 10^0 = \boldsymbol{D}_b \cdot 10 + \boldsymbol{U}_b \\
\boldsymbol{c} \cdot n &= \boldsymbol{D}_c \cdot 10^1 + \boldsymbol{U}_c \cdot 10^0 = \boldsymbol{D}_c \cdot 10 + \boldsymbol{U}_c
\end{aligned}$$

where D_a, D_b and D_c are respectively the tens of the number resulting from multiplying n by a, b, or c, and U_a, U_b and U_c are their respective units.

By multiplying cba times n, we are doing the following operations:

$$\begin{aligned}
\boldsymbol{cba} \cdot n &= (\boldsymbol{c} \cdot 100 + \boldsymbol{b} \cdot 10 + \boldsymbol{a}) \cdot n \\
&= (\boldsymbol{c} \cdot n) \cdot 100 + (\boldsymbol{b} \cdot n) \cdot 10 + (\boldsymbol{a} \cdot n) \\
&= (\boldsymbol{D}_c \cdot 10 + \boldsymbol{U}_c) \cdot 100 + (\boldsymbol{D}_b \cdot 10 + \boldsymbol{U}_b) \cdot 10 \\
&\quad + (\boldsymbol{D}_a \cdot 10 + \boldsymbol{U}_a) \\
&= \boldsymbol{D}_c \cdot 1000 + \boldsymbol{U}_c \cdot 100 + \boldsymbol{D}_b \cdot 100 + \boldsymbol{U}_b \cdot 10 \\
&\quad + \boldsymbol{D}_a \cdot 10 + \boldsymbol{U}_a
\end{aligned}$$

and grouping terms:

$$\boldsymbol{cba} \cdot n = \boldsymbol{D}_c \cdot 10^3 + (\boldsymbol{U}_c + \boldsymbol{D}_b) \cdot 10^2 + (\boldsymbol{U}_b + \boldsymbol{D}_a) \cdot 10 + \boldsymbol{U}_a$$

that is, the resulting number is such that:

- The **units** of cba are $\boldsymbol{U}_a$ (the units of a).
- The **tens** of cba are $\boldsymbol{U}_b + \boldsymbol{D}_a$ (the sum of the units of b plus the tens of a).
- The **hundreds** of cba are $\boldsymbol{U}_c + \boldsymbol{D}_b$ (the sum of the units of c, the tens of b and of course the carry generated by $\boldsymbol{U}_b + \boldsymbol{D}_a$).
- The **thousands** of cba are $\boldsymbol{D}_c$ (the tens of c, and the immediately preceding carry).

guidelines that summarize the algorithm that we are going to use next.

Single-digit multiplier

The multiplication by a single digit is easy. The pattern is always ***UD** moving from right to left over the multiplicand, with **U** coinciding over the same position as the digit of the result being calculated. Explicitly, the patterns to get each digit of the result (from right to left) are:* $\boldsymbol{U}$, $\boldsymbol{UD}$, $\boldsymbol{UDo}$,...

Let's see *for example* how to multiply $\mathbf{4329 \cdot 8}$, whose result is **34 632**. We consider, as usual, the multiplier preceded by as many zeros as figures there are in the multiplier (**04329** in this case):

For the **units of the result**, the *pattern* is ***U***, which only affects the units of the multiplicand, namely, **9**. We must multiply 9 *by* 8 (which is 72) and take only the digit corresponding to the units of the product, that is, **2**.

For the **tens of the result**, the *pattern* is ***UD***, in which ***U** affects the tens of the multiplicand* (**2**) and ***D** affects the units* of the same (**9**). We add the units (*U pattern*) of the product 2 *times* 8 (the digit **6** of **16**) to the tens (*D pattern*) of the product 9 *times* 8 (the digit **7** of **72**); the result is **13** (**3** units and 1 *carry*). *The addends generated by the pattern correspond to the central digits (6 and 7) of the two products, when they are placed one adjacent to the other (16 72).*

For the **hundreds of the result**, the *pattern* is ***UDo***, in which ***U** affects the hundreds* of the multiplicand (**3**) and ***D** affects the tens* of the same (**2**). We add the units (*U pattern*) of the product 3 *times* 8 (the digit **4** of **24**) to the tens (*D pattern*) of the product 2 *times* 8 (the digit **1** of **16**), which added to the previous carry is **6** (4 plus 1 is **5**, plus **1** *carry* makes 6).

For the **units of thousands of the result**, the

pattern is ***UDoo***, in which ***U*** *affects the thousand units* of the multiplicand (**4**) and ***D*** *affects the hundreds* of it (**3**). We add the units (*U pattern*) of the product 4 *times* 8 (the digit **2** of 3**2**) to the tens (*D pattern*) of the product 3 *times* 8 (the digit **2** of **2**4), and since there is no carry, the result is **4** (2 plus 2).

For the **tens of thousands of the result**, the *pattern* is ***UDooo***, in which ***U*** *affects the tens of thousands* of the multiplicand (**0**) and ***D*** *affects the units of thousands* of the same (**4**). We add the units (*U pattern*) of the product 0 *times* 8 (the digit **0** of 0**0**) to the tens (*D pattern*) of the product 4 *times* 8 (the digit **3** of **3**2), and because there is no carry, the result is **3** (0 plus 3).

Two-digit multiplier

In the multiplication by two figures, there are going to be two patterns for each digit of the result, the one associated with the tens of the multiplier is similar to the one used in the previous point and the one corresponding to the units of the multiplier, is the same one displaced one position to the left relative to this (if that one was *UDo*, this one is *UDoo*). This produces four factors that, added to a possible previous carry, generate the desired digit of the result. These two patterns must shift from right to left as in single-digit multiplier multiplication. *The U of the longest pattern must always coincide in position with that of the digit of the result to be calculated.*

Let's see *for example* how to multiply **6543 · 21**, whose result is **137 403**. We precede the multiplicand with as many zeros as there are digits in the multiplier, which in this case are two (**006543**):

For the **units of the result**, the digit **3** of the multiplicand is needed; only the pattern corresponding to the units of the multiplier (***U***) is required, which only affects the units of the multiplicand, namely **3**. We have to multiply 3 *by* 1 (which is 0**3**) and take only the digit corresponding to the units of that product, that is, **3**.

For the **tens of the result**, the digits **43** of the multiplicand are needed; two addends are obtained through the ***UD*** pattern *(corresponding to the units of the multiplier)* in which ***U*** *affects the tens* of the multiplicand (**4**) and ***D*** *affects the units* of the same (**3**); the other uses the ***U*** pattern *(corresponding to the tens of the multiplier) affecting the units* of the multiplicand

(**3**). The three products are respectively **04** (4 *times* 1), **03** (3 *times* 1) and **06** (3 *times* 2), which, applying the corresponding pattern, produce the addends: **4** (*pattern* ***U*** **over 04**), **0** (*pattern* ***D*** **over 03**) and **6** (*pattern* ***U*** **over 06**) and added together are **10** (**0** units and 1 *carry*).

For the **hundreds of the result**, the digits **543** of the multiplicand are needed; 2 addends are obtained through the ***UDo*** *pattern (corresponding to the units of the multiplier)* in which ***U*** *affects the hundreds* of the multiplicand (**5**) and ***D*** *to the tens of the same* (**4**); the other two are generated with the ***UD*** *pattern (corresponding to the tens of the multiplier)* in which ***U*** *affects the tens* of the multiplicand (**4**) and ***D*** *affects the units* of the same (**3**). The four products are respectively **05** (5 *times* 1), **04** (4 *times* 1), **08** (4 *times* 2) and **06** (3 *times* 2), which, applying the corresponding pattern, produce the addends **5** (*pattern* ***U*** **over** 05), **0** (*pattern* ***D*** **over** 04), **8** (*pattern* ***U*** **over** 08) and **0** (*pattern* ***D*** **over** 06) which together with the previous carry make a total of **14** (**4** units and 1 *carry*).

For the **units of thousands of the result**, the digits **6543** of the multiplicand are needed, but only the leftmost three are used in the calculation; two addends are obtained through the ***UDoo** pattern (corresponding to the units of the multiplier)* in which ***U** affects the units of thousands* of the multiplicand (**6**) and ***D** affects the hundreds* of the same (**5**); the other two are generated with the ***UDo** pattern (corresponding to the tens of the multiplier)* in which ***U** affects the hundreds* of the multiplicand (**5**) and ***D** affects the tens* of the same (**4**). The four products are respectively **06** (6 *times* 1), **05** (5 *times* 1), **10** (5 *times* 2) and **08** (4 *times* 2), which, applying the corresponding pattern, produce the addends **6** (*pattern* ***U*** **over** 0**6**), **0** (*pattern* ***D*** **over** **0**5), **0** (*pattern* ***U*** **over** 1**0**) and **0** (*pattern* ***D*** **over** **0**8), and added to the previous carry make a total of **7**.

For the **tens of thousands of the result**, the digits **06543** of the multiplicand are needed, but only the leftmost three are used in the calculation; two addends are obtained using the ***UDooo** pattern (corresponding to the units of the multiplier)* in which ***U** affects the tens of thousands* of the multiplicand (**0**) and ***D** affects the units of thousands* of the same (**6**); the other two are generated with the ***UDoo** pattern (corresponding to the tens of the multiplier)* in which ***U*** *affects the units of thousands* of the multiplicand (**6**) and ***D** to the hundreds* of the same (**5**). The four products are respectively **00** (0 *times* 1), **06** (6 *times* 1), **12** (6 *times* 2) and **10** (5 *times* 2), which by applying the corresponding pattern determine the addends **0** (*pattern* ***U*** **over** 00), **0** (*pattern* ***D*** **over** 06), **2** (*pattern* ***U*** **over** 12) and **1** (*pattern* ***D*** **over** 10) whose sum is **3**.

For the **hundreds of thousands of the result**, the digits **006543** of the multiplicand are needed, but only the leftmost three are used in the calculation; two addends are obtained using the ***UDoooo** pattern (corresponding to the units of the multiplier)* in which ***U** affects the hundreds of thousands* of the multiplicand (**0**) and ***D** affects the tens of thousands of the same* (**0**); the other two are generated with the ***UDooo** pattern (corresponding to the tens of the multiplier)* in which ***U** affects the tens of thousands* of the multiplicand (**0**) and ***D** to the units of thousands* of the same (**6**). The four products are respectively **00** (0 *times* 1), **00** (0 *times* 1), **00** (0 *times* 2) and **12** (6 *times* 2); that applying the corresponding pattern determine the addends **0** (*pattern* ***U*** **over** 00), **0** (*pattern* ***D*** **over** 00), **2** (*pattern* ***U*** **over** 00) and **1** (*pattern* ***D*** **over** 12) whose sum makes a total of **1**.

In general, the patterns act on three digits of the multiplicand. If we consider that these are **543** and the *pattern associated with the units of the multiplier* is ***UDo*** (acting over the **5** and **4**), necessarily the *pattern corresponding to the tens of the multiplier* is ***UD*** (acting over the **4** and **3**). If the multiplier is **21** then to find the hundreds digit of the result we calculate 5 *times* 1, 4 *times* 1, 4 *times* 2 and 3 *times* 2, applying the ***UDUD*** pattern respectively; the resulting four numbers are added with the previous carry (if any) producing a digit and a possible carry.

Three or more digits multiplier

As will be seen, the procedure to follow is similar to the previous one. There is a *first **UD** **pattern** associated with the units of the multiplier so that **U*** operates with the digit of the multiplicand located in the ***P position,*** the same as the digit of the result to be calculated; ***D*** operates over the digit of the multiplicand in *position **P-1*** (if possible). Un *second **UD** pattern associated with the tens of the multiplie*r operates with the multiplicand digits located in positions $P-1$ and $P-2$ (if that is possible) and a *third UD pattern associated with the hundreds of the multiplier* operates with the digits of the multiplicand located in positions $\boldsymbol{P-2}$ and $\boldsymbol{P-3}$ (if possible). *Each digit of the multiplier is multiplied by the digits of the multiplicand affected by the pattern, choosing the units or tens* ***units or tens*** *of each operation as appropriate (**U or D**); in this way six numbers are obtained (two per pattern) that together with the previous carry define the sought digit of the result.*

(We consider that positions increase from right to left so that the *P position* is one place further to the left than *position* $P-1$).

Adding one more digit to the multiplier means **generating another *UD pattern*** associated with the new digit following the above rules, that is, the **pattern associated with the units of the multiplier** defines the **positions** $\boldsymbol{P}$ and $\boldsymbol{P-1}$; the **rest** will define **positions** $\boldsymbol{P-1}$ and $\boldsymbol{P-2}$, $\boldsymbol{P-2}$ and $\boldsymbol{P-3}$, etc. until there are no more unused digits in the multiplier. The way of operating is the same; only the number of addends per digit to calculate varies.

Let's see *for example* how to multiply $\mathbf{7654 \cdot 321}$, whose result is **2 456 934**. We consider, as usual, the multiplier preceded by as many zeros as figures there are in the multiplier (**0007654** in this case):

To find the **units of the result**, the pattern that corresponds to the units of the multiplier is ***U***, which is applied to the product 4 *times* 1 (of the multiplicand and the multiplier, respectively) resulting in **4** (that is, the units of 0**4**).

For the **tens of the result**, he decisive part of the multiplicand is **54**. The pattern that corresponds to the units of the multiplier (**1**) is ***UD*** *(where U is in the same position as 5);* the units of 0**5** (5 *times* 1 is 0**5**) are **5** y and the tens of **0**4 (4 *times* 1 is **0**4) are **0**. The pattern that corresponds to the tens of the multiplier (**2**) is ***U*** *(where U is in the same position as 4);* the units of 0**8** (4 *times* 2) are **8**. There are only three addends (**5**, **0**, and **8**) and their sum is **13** (**3** units and 1 *carry*).

For the **hundreds of the result**, the necessary part of the multiplicand is **654**. The pattern that corresponds to the units of the multiplier (**1**) is ***UD*** *(where U is in the same position as 6);* the units of 0**6** (6 *times* 1) are **6** and tens of **0**5 (5 *times* 1) are **0**. The pattern corresponding to the tens of the multiplier (**2**) is

UD *(where U is in the same position as 5);* the units of **10** (5 *times* 2) are **0** and the tens of **0**8 (4 *times* 2 are **08**) are **0**. The pattern that corresponds to the hundreds of the multiplier (**3**) is ***U*** *(where U is in the same position than 4);* the units of **12** (4 *times* 3) are **2**. There are only the five addends **6**, **0**, **0**, **0** and **2** that together with **1** *carry* add up to **9**.

For the **units of thousands of the result**, the important part of the multiplicand is **7654**. The pattern that corresponds to the units of the multiplier (**1**) is ***UD*** *(where U is in the same position as 7);* the units of **07** (7 *times* 1) are **7**, and the tens of **06** (6 *times* 1) are **0**. The pattern associated with the tens of the multiplier (**2**) is ***UD*** *(where U is in the same position as 6);* the units of **12** (6 *times* 2) are **2**, and the tens of **10** (5 *times* 2) are **1**. The pattern associated with the hundreds of the multiplier (**3**) is ***UD*** *(where U is in the same position as 5);* the units of **15** (5 *times* 3) are **5** and the tens of **12** (4 *times* 3) are **1**. We already have the six addends **7**, **0**, **2**, **1**, **5** and **1** which add up to **16** (**6** units and 1 *carry*).

For the **tens of thousands of the result**, the important part of the multiplicand is **0765**. The pattern that corresponds to the units of the multiplier (**1**) is ***UD*** *(where U is in the same position as 0);* the units of 0**0** (0 *times* 1) are **0** and the tens of **0**7 (7 *times* 1) are **0**. The pattern associated with the tens of the multiplier (**2**) is ***UD*** *(where U is in the same position as 7);* the units of 1**4** (7 *times* 2) are **4** and the tens of **1**2 (6 *times* 2) are **1**. The pattern associated with the hundreds of the multiplier (**3**) is ***UD*** *(where U is in the same position as 6);* the units of 1**8** (6 *times* 3) are **8** and the tens of **1**5 (5 *times* 3) are **1**. Gathering information, the six addends are **0**, **0**, **4**, **1**, **8** and **1**, which plus 1 *carry* equals **15** (**5** units and 1 *carry*).

For the **hundreds of thousands of the result**, the important part of the multiplicand is **0076**. The pattern that corresponds to the units of the multiplier (**1**) is ***UD*** *(where U is in the same position as the first 0 from the left);* the units of 0**0** (0 *times* 1) are **0** and the tens of **0**0 (0 *times* 1) are **0**. The pattern associated with the tens of the ***2*** multiplier is ***UD*** *(with U in the same position as the second 0 from the left);* the units of 0**0** (0 *times* 2) are **0** and the tens of **1**4 (7 *times* 2) are **1**. The pattern associated with the hundreds of the multiplier (**3**) is ***UD*** *(where U is in the same position as 7);* the units of 2**1** (7 *times* 3) are **1** and the tens of **1**8 (6 *times* 3) are **1**. We already have the six addends **0**, **0**, **0**, **1**, **1** and **1**, that added to 1 from the previous *carry* add up to **4**.

For the **units of millions of the result**, the important part of the multiplicand is **0007**. The pattern that corresponds to the units of the multiplier (**1**) is ***UD*** *(where U is in the same position as the first 0 from the left);* this generates two zeros. The same happens with the pattern associated with the tens of the multiplier (**2**) ***UD*** *(where U is in the same position as the second 0).* The pattern associated with the hundreds of the multiplier (**3**) is ***UD*** *(where U is in the same position as the third 0 from the left);* the units of 0**0** (0 *times* 3) are **0** and the tens of **2**1 (7 *times* 3) are **2**. Finally, we have the six addends **0**, **0**, **0**, **0**, **0** and **2** whose sum is **2**.

Chapter 4

Additions and corrections

In this chapter we will focus on how to easily perform long sums of many numbers, on how to find errors when performing the calculation, and on narrowing down those errors so that we do not have to check the entire sum but only part of it.

In the traditional way, the units of each addend are added, obtaining a result and a possible carry that is added to the sum of the tens of each addend to produce another result and another carry, which is added to the sum of the hundreds to produce a third result and consequent carry... and so on until end up with the numbers from right to left.

The carry occurs whenever a unit is added to 9 to produce 10. One way to ease the calculation is precisely to count how many times this number has been exceeded. To do this, an imaginary counter is incremented each time this happens and the addition is continued with the units of the last result, ignoring the tens (that is, we subtract 10). Each increment of the tens counter is a carry to be accounted for in the next column to the left. For example, let's consider the sum of the two-digit numbers **64**, **25**, **17**, **58**, **43**, **93** and **7**; from top to bottom the *units of each addend* of the operation are **4**, **5**, **7**, **8**, **3**, **3** and **7**. We begin by adding **4** and **5** (which are 9); 9 and **7** count as 6 (because they are 16 and the 1 is considered but *ignored*); 6 and **8** produce 4 (because they add up to 14); 4 and **3** are 7; 7 and **3** generate 0 (because their sum is 10); and finally, 0 and **7** are 7, **determining** the **units** of the result **in 7** and the **counter of tens of the units** in **3**, which, although it does not influence the units, does influence the tens; traditionally we get a total of 37 (7 units and 3 *carry*). Now it is the turn of *the tens of each addend*, these are **6**, **2**, **1**, **5**, **4**, **9** and **0**. We proceed with the addition, as before; **6** and **2** are 8; 8 and **1** are 9; 9 and **5** make 14; 4 and **4** are 8; 8 and **9** are 17; and the last one, 7 and **0** are 7, which derives in 7 and the **tens counter of the tens** in **2**; *the tens* of the result are then

7 plus **3** (the counter of tens of the units), that is, **10** (**0** and 1 *carry* that is transferred as the sum of the following column); the hundreds of the result are made up of the sum of the tens counter (**2**) with the last carry (**1**). The total sum is therefore **307**, which coincides with the result provided by the traditional computation method (the sum of the tens 6, 2, 1, 5, 4, 9 and 0 is 27 that added to 3 of the previous carry generated by the sum of the units is 30).

We can do the same example *by now considering* **counting the number of elevens** instead of the number of tens. The process is the same except that each time the digit 11 is exceeded, instead of simply ignoring the tens for the calculation (although they count for the number of elevens), the units must be decremented by one unit (which is equivalent to *subtracting* 11). This fact makes it necessary to replace as many ones as elevens have been found in the column that is being added (whether they are units, tens, etc.), which at first may seem like a disadvantage; however, this method makes it easier to correct errors as we will see later in this chapter. Let's proceed with the example; the addends are the same: **64**, **25**, **17**, **58**, **43**, **93** and **7**. As before, *each unit of each addend* (**4**, **5**, **7**, **8**, **3**, **3** and **7**) is considered first (from top to bottom, although another order could be followed) by adding them as follows: **4** and **5** are 9; 9 and **7** are 16 (the 1 is ignored, *counting it for the elevens' counter* of the units —now 1— and 6 is decremented by 1); 5 and **8** are 13 (the elevens' counter of the units is increased again —now 2— and 3 is decreased by 1); 2 and **3** are 5; 5 and **3** are 8; and finally, 8 and **7** are 15 (new increment of the counter —which is now 3— and decrement of 5 by 1), which results in **4** and an **elevens' counter for the units** of **3** (quantity that replenishes as many ones as elevens we had subtracted) that added (**7**) determine the **units** for the final **result**. We now consider *each ten of each addend* (in the same order as before), namely, **6**, **2**, **1**, **5**, **4**, **9** and **0**. The addition process goes like this: 6 and **2** are 8; 8 and **1** are 9; 9 and **5** are 14 (so the *elevens' counter for the tens* is now **1**, and 4 must be decremented by 1); 3 and **4** are 7; 7 and **9** are 16 (new increment in the **elevens' counter of the tens** —which is now **2**— and decrement of 6 by one unit); and finally, 5 and **0** are **5**. To determine the tens of the result, we must add to this **5** as many ones as elevens we had subtracted in the tens column (the **elevens' counter of**

the tens is **2**) and the previous carry associated with the subtraction of numbers 11 (the ***elevens' counter of the units*** *is* **3**), which adds up to **10** (**0** units and 1 *carry* of the result to be added to the next column on the left). In the hundreds of the final result there are no ones to replace (no 11 has been subtracted) but it is appropriate to add the sum of the last carry generated in the previous column (which is **1**) and the carry that carries with it the previous elevens' counter (that of the tens, which is **2**), so the hundreds of the result are **3** (1 plus 2) and the final result of the addition is **307**.

The relevant data of the calculation process is condensed in the following table, where the number of asterisk characters (*) to the right of a number indicates the amount of the carry:

	Hundreds	*Tens*	*Units*
Total calculation	0	5	4
Elevens' counter	0	2	3
Operation result	3	0*	7

Each digit of the result is found by adding the elevens' counter of the same column and the previous one to the total calculation of the corresponding column, that is: the units are **7** (4 plus 3), the tens are **0***(5 plus 2 plus 3, which adds up to 10 —0 units and a carry of 1, denoted by an asterisk to the right of the 0—) and hundreds are **3** (0 plus 0 plus 2 plus 1 *carry* that was generated in the previous column).

The **addition of numbers with decimals** is treated in the same way; the only thing essential is to position each digit in the appropriate column according to its weighting (to the left of the decimal point, units with units, tens with tens, hundreds with hundreds, etc. and to the right of said point, tenths with tenths, hundredths with hundredths, thousandths with thousandths, etc.). It should be clarified that the decimal separator that we will use is the point (as is common practice in English). To add 0.69 plus 3.28 plus 0.75 we proceed by placing each digit in its corresponding column. From top to bottom and from right to left: **9**, **8** and **5** belong to the hundredths' column; **6**, **2** and **7** are part of the tenths; and **0**, **0** and **3** correspond to the units. If we put in parentheses each increment of the elevens' counter (which is initialized to 0 in each column) the calculation (from right to left, as usual) would be: **9** plus **8** are 17, so, we decrement 7 by 1 and increment the elevens' counter for the column, that is, **6**(**1**); 6 and **5** are 11 ending the column with the decrement of 1 by one unit and the increment of the elevens' counter by 1, that is, **0**(**2**). Following the same format in the tenths column we have: **6** and **2** are 8; 8 and **7** are 15 —**4**(**1**). The ones column yields only **3**(**0**).

	Units	Tenths	Hundredths
	0.	6	9
	3.	2	8
	0.	7	5
Total calculation	3	**4**	0
Elevens' counter	0	**1**	**2**
(Operation detail)	$3 + 0 + 1$	$\mathbf{4 + 1 + 2}$	$0 + 2$
Operation result	4.	7	2

For clarity, a row has been included indicating the details of each final operation and the digits that participate in the calculation of the tenths of the result have been highlighted in bold.

The first digits of the number π added in a certain

way provide the curious result 69 999 (a number that after subtracting 7 and dividing it by 13 gives exactly 5384). Let's do this example.

The thirteen addends are: **31415**, **9265**, **3589**, **793**, **2384**, **626**, **4338**, **3279**, **5028**, **8419**, **71**, **693** and **99**.

The **sum of the units** of each addend is (from first to last): **5** plus **5** are 10; 10 and **9** are 19 «8(1)»; 8 and **3** are 11 «0(2)»; 0 and **4** are 4; 4 and **6** are 10; 10 and **8** are 18 «7(3)»; 7 and **9** are 16 «5(4)»; 5 and **8** are 13 «2(5)»; 2 and **9** are 11 «0(6)»; 0 and **1** is 1; 1 and **3** are 4; and finally, 4 and **9** are 13 «**2**(**7**)», that is, the total calculation of the *units* is **2** and the *elevens' counter* of the *units* is **7**; their sum (**9**) establishes the units of the final result.

The **sum of the tens** of each addend is (from first to last): **1** and **6** are 7; 7 and **8** are 15 «4(1)»; 4 and **9** are 13 «2(2)»; 2 plus **8** is 10; 10 plus **2** is 12 «1(3)»; 1 and **3** are 4; 4 and **7** are 11 «0(4)» 0 and **2** are 2; 2 and **1** are 3; 3 and **7** are 10; 10 and **9** are 19 «8(5)»; and finally, 8 and **9** are 17 «**6**(**6**)», that is, a *total calculation of tens of **6** and an elevens' counter for the tens of **6**;* the **tens of the result** are determined by adding to these numbers (6 and 6 are **12**) the elevens' counter of the units (**7**), a total of **19** (**9** units and 1 *carry*, which with the notation used above we can represent as **9***).

The **sum of the hundreds** of each addend is (from first to last): **4** and **2** are 6; 6 and **5** are 11 «0(1)»; 0 and **7** are 7; 7 and **3** are 10; 10 and **6** are 16 «5(2)»; 5 and **3** are 8; 8 and **2** are 10; 10 and **0** are 10 ; 10 and **4** are 14 «3(3)»; and finally 3 and **6** are 9 — in short, **9**(**3**)—, that is, *total calculation of hundreds **9** and counter of elevens of the hundreds **3**;* the hundreds of the result are determined by adding to these two numbers (9 and 3 are **12**) the elevens' counter of the tens (**6**) and the general carry (**1**), which totals **19** (**9** units and a carry of 1, which again we can represent as **9***).

The **sum of the units of thousands** of each addend is (from the first to the last): **1** and **9** are 10; 10 and **3** are 13 «2(1)»; 2 and **2** are 4; 4 and **4** are 8; 8 and **3** are 11 «0(2)»; 0 and **5** are 5; and finally, 5 and **8** are 13 «**2**(**3**)», that is, *total calculation of units of thousands **2** and counter of elevens of the units of thousands **3**;* the units of thousands of the result are determined by adding to these two numbers (2 and 3 are **5**) the counter of elevens of the hundreds (**3**) and the immediately previous general carry (**1**), which totals **9**.

The **sum of the tens of thousands** of each addend is (from first to last): **3** (the rest are zero). The elevens' counter of the tens of thousands is 0, the elevens' counter of the units of thousands is **3** and there is no previous general carry; so, the tens of thousands of the result are **6** (3 plus 3).

As we have seen with the previous example, the calculation is simple; we only have to consider the last sum made, the elevens' counters (previous and current) and the possible general carry. The following table summarizes the most relevant data (where ***U*** are the units, ***T*** the tens, ***H*** the hundreds, ***UT*** the units of thousands and ***TT*** the tens of thousands):

	TT	***UT***	***H***	***T***	***U***
Total calculation	3	2	9	6	2
Elevens' counter	0	3	3	6	7
Operation result	6	9	9*	9*	9

Detection of errors in operations

To check the operations, it's necessary to obtain a check

digit from each column (units, tens, etc.), several check digits (also one per column) from what we are going to designate from now ***work table*** composed of the first two lines of the last table of the previous point (identified as *«Total calculation»* y *«Elevens' counter»*), and another control digit obtained from the result of the operation.

We use the same example as in the previous point, that is, the thirteen addends: **31415**, **9265**, **3589**, **793**, **2384**, **626**, **4338**, **3279**, **5028**, **8419**, **71**, **693** and **99**.

We proceed with the calculation of the **check digit of each column**. For this we must ignore all the nines (also the zeros) and the digits that added are 9 or a multiple of 9 of the target column, to then add the rest and thus obtain a number whose figures must be reduced to a single digit by adding them (if applicable):

The digits in the **units' column** are: **5**, **5**, **9**, **3**, **4**, **6**, **8**, **9**, **8**, **9**, **1**, **3** and **9**. We can remove four nines, and from the rest, those that add up to 9, namely, «**5** and **4**», «**6** and **3**», «**8** and **1**», leaving only **5**, **8** and **3** that *generate the digit* **7** (5 and 8 are 13; 1 and 3 are 4; 4 and 3 are **7** —or 5 plus 8 plus 3 equals 16; 1 plus 6 equals **7**).

The digits of the **tens' column** are: **1**, **6**, **8**, **9**, **8**, **2**, **3**, **7**, **2**, **1**, **7**, **9** and **9**. We can eliminate three nines, and from the rest, those that add up to 9, namely (taken from left to right), «**1** and **8**», «**6** and **3**», «**8** and **1**», «**2** and **7**», «**2** and **7**», completely using all the numbers and thus *generating the digit* **0** (which for practical purposes is equivalent to **9**).

The digits in the **hundreds' column** are: **4**, **2**, **5**, **7**, **3**, **6**, **3**, **2**, **0**, **4** and **6**. There are no nines to remove, although the **0** can be ignored, and from the rest, we can remove those that add up to 9, namely, «**4** plus **2** plus **3**», «**5** and **4**», «**7** and **2**», «**6** and **3**», leaving only **6**, which becomes the *check digit*.

The digits of the **units of thousands** column are: **1**, **9**, **3**, **2**, **4**, **3**, **5** and **8**. We can eliminate a nine, and from the rest, those that add up to 9, namely (taken from left to right), «**1** and **8**», «**3** plus **2** plus **4**», which defines the *check digit* as the sum of «**3** and **5**».

As for the **tens of thousands** column, the only digit available is **3**, which is itself the *check digit* for this column.

The following table summarizes the **check digits** obtained for each column (***U*** units, ***T*** tens, ***H*** hundreds, ***UT*** units of thousands and ***TT*** tens of thousands):

	TT	***UT***	***H***	***T***	***U***
Check digits	3	8	6	0 *or* 9	7

The following check digits are obtained from the table that we have called above **work table**, but *by duplicating the line of the elevens' counter* before proceeding with the addition of each column, that is:

	TT	***UT***	***H***	***T***	***U***
Total calculation	3	2	9	6	2
Elevens' counter	0	3	3	6	7
Elevens' counter	0	3	3	6	7
Sum of the column	3	8	15	18	16
Check digits	**3**	**8**	**6**	**9**	**7**

The **units' check digit** comes from the sum of the elements of column ***U*** through the sequence of operations: **2** plus **7** plus **7** equals 16; 1 and 6 are **7**.

The **tens' check digit** comes from adding the

elements of column ***T*** using the sequence of operations: **6** plus **6** plus **6** equals 18; 1 and 8 are **9**.

The **hundreds' check digit** comes from the sum of the elements of column ***H*** through the sequence of operations: **9** plus **3** plus **3** are 15; 1 and 5 are **6**.

The **check digit of the units of thousands** comes from the sum of the elements of the ***UT*** column through the operation: **2** plus **3** plus **3** equals **8**.

The **tens of thousands check digit** comes from the sum of the elements of the ***TT*** column using the operation: **3** plus **0** plus **0** equals **3**.

The result is identical to the calculation of the first check digits obtained above; this indicates that there is no inconsistency between the total calculation and the amount of eleven numbers counted in any of the columns.

The last check digit is obtained from the row corresponding to the result of the operation, in this case:

	TT	***UT***	***H***	***T***	***U***
Operation result	6	9	9	9	9

Here we only have to reduce all the figures that are part of the result to just one by successive addition of its digits. The conversion to a single digit can be done at the end of the sum or at each partial sum, that is, we can add **6** plus **9** plus **9** plus **9** plus **9** (42 total) and then 4 plus 2 to get the *check digit* **6**, or else or carry out a sequence of operations similar to the following: **6** and **9** are 15; 1 and 5 are 6; 6 and **9** are 15; 1 and 5 are 6; and so on until we get the check digit (**6**).

Previously we obtained as check digits of each column: **3**, **8**, **6**, **9** and **7**. We can determine a single digit from them by adding them and simplifying the result as before: **3** and **8** are 11; 1 and 1 are 2; 2 and **6** are 8; 8 and **9** are 17; 1 and 7 are 8; 8 y **7** son 15; and finally, 1 and 5 are **6**, a value that coincides with the check digit found for the result of the operation, which is indicative that everything is correct.

General checking methods

We are going to look at two essential methods for checking errors in an operation: the *reduction to digits method* and the *elevens' method*.

Reduction to digits method

This method consists of reducing each component of the operation to a single digit, performing said operation with those digits (addition, subtraction, multiplication or division) and verifying that the resulting digit matches the one found for the result. **If it does NOT match, an error has been made** when operating.

Reducing to a single digit is simple. As before, we can add all the digits of the number and reduce the final result to a single digit by repeating the same process (as in the number 856 —8 plus 5 plus 6 is 19; 1 and 9 are 10; 1 and 0 is **1**) or transform each partial result to a single digit (again using 856 as an example, 8 and 5 are 13; 1 and 3 are 4; 4 and 6 are 10; 1 and 0 is **1**).

Also, there is an interesting property that makes the computation easier: **9*s* and 0*s* can be ignored**. *Adding one or more nines (this includes all multiples of 9) to the resulting digit does not affect the final result.* Adding 9 is adding 10 and subtracting 1, but ten reduced to a single digit is 1 (1 and 0 is 1), so we're actually adding 0. Since 9 behaves like 0 we can say that *0 and 9 can be considered the same* when deciding whether a result is correct or not.

The above rule is really powerful:

All digits of the number that together add up to 9 can be ignored. E.g., **136 847 299** is easily reduced to **4** since «**1** and **8**» are 9; «**3** and **6**» are 9; «**7** and **2**» are 9; and finally, the two nines can be ignored leaving only the digit **4**. The same calculation done as before would be more laborious: **1** and **3** are 4; 4 and **6** are 1 (1 plus 0); 1 and **8** are 9; 9 and **4** are 4 (1 plus 3); 4 and **7** are 2 (1 plus 1); 2 and **2** are 4; 4 and **9** are 4 (1 plus 3); and finally, 4 and **9** are again **4**.

There are times that *the process of reducing the number to a single digit can be simplified by taking the remainder of the division of said number by 9.* For example, the number **2747** when divided by 9 gives a remainder of **2**, the same digit that results when reducing 2747 to just one digit: since 2 and 7 add up to 9, they are ignored; 7 is equal to 5 plus 2, this allows us to ignore 4 and 5 — because they add up to 9— leaving only **2** as a result.

All this helps to **detect that an operation is incorrect**, but the correctness of the operation is not guaranteed, as we can see in the following example.

The sum of **456** plus **21** is **477**. Reducing each

addend to a single digit, 456 is **6** and 21 is **3**. The sum of 6 and 3 (which is **9**) gives the same digit as the one obtained from the result (**477** *is* 9 as a single digit) so the operation seems to be correct (**387** *is also* ***9*** reduced to a single digit *even if it is a* **wrong answer**). However, *when faced with a result like* **487**, we can say without a doubt that the operation **is incorrect**, since 487 is **1** reduced to a single digit.

If we say that the product of **12** and **4** is **41**, *this error **can be detected**.* The first factor becomes **3**, the second is **4**; their product is **12**, which as a digit is **3**. On the other hand, 41 as a digit is $\mathbf{5} \neq \mathbf{3}$, a discrepancy that alerts to a failure in the calculation.

The elevens' method

This method consists of reducing each component of the operation to a number between zero and ten (by calculating the remainder by dividing by eleven using addition and subtraction), performing said operation with those digits (addition, subtraction, multiplication or division) and check that the resulting digit matches the one obtained for the result. **If it does NOT match, an error has been made** when operating.

The calculation of **the remainder when dividing** a **two-digit number by 11** is quite simple, the maximum multiple of eleven that the number can contain is sought and subtracted from it. But to follow an analogy with the algorithm that we are going to use when the number has more figures, we are going to use this other procedure (even simpler):

If the **units** are **greater than** the **tens**, we only have to *subtract the tens' digit from the units' digit*; the result is the remainder when divided by 11. For example, if the target number is 78, the remainder when divided by 11 is **1** (8 minus 7), which is true, since 78 is 7 times 11 plus **1**.

If the **units** are **less than** the **tens**, we have to *add 11 to the units before subtracting it the tens' digit;* the result is the remainder when divided by 11. For example, if the target number is 82, we must add 11 to 2 (that is, 13) and subtract 8 to obtain the remainder when divided by 11; since 13 minus 8 is **5**, this is the remainder we are looking for, which is true, since 82 is 7 times 11 plus **5**.

The calculation of the **remainder when dividing by 11** a **number of more than two figures** is more laborious, but applying the method indicated in the two previous paragraphs on *certain sets* of numbers, the computation is reduced to simple additions and subtractions:

The **first set of numbers** to consider is the *units' digit* itself and *every second digit* of all those to its left.

The **second set of numbers** is the *tens' digit* itself and *every second digit* of all those to its left.

To find the remainder when dividing by 11, we just need *to subtract the sum of the first set of numbers from the sum of the second set of numbers and reduce the result to a multiple of 11.* Of special importance is that if *the numerator of this subtraction is less than the denominator*, we must add 11 to the numerator before proceeding with the subtraction.

For example, let's consider the 20-digit number 9**3** 4**5**2 **6**7**8** 3**4**5 **6**7**8** 9**9**8 **8**2**1** (whose remainder *when divided by 11* is **2**, since it is 8 495 698 031 425 363 529 times 11 plus **2**).

The first set of numbers is **1**, **8**, **9**, **8**, **6**, **4**, **8**, **6**, **5** and **3**; their sum is **58**.

The second set of numbers is **2**, **8**, **9**, **7**, **5**, **3**, **7**, **2**,

4 and **9**; their sum is **56**.

The **subtraction 58 minus 56** gives us *the remainder when dividing by the* desired *11*, namely **2**. We could also have reduced **58** and **56** to their *respective remainders by dividing by 11*, that is, **3** (8 minus 5) and **1** (6 minus 5) and subtract 3 minus 1 to get the same result.

Equally valid would have been:

First, reduce each sum in the first group to the remainder when dividing by 11, like this: **1** and **8** are 9; 9 and **9** are 18; «8 minus 1 is 7» *(to find the remainder);* 7 and **8** are 15; «5 minus 1 is 4»; 4 and **6** are 10; 10 and **4** are 14; «4 minus 1 is 3»; 3 and **8** are 11; «1 minus 1 is 0»; 0 and **6** are 6; 6 and **5** are 11; «1 minus 1 is 0»; and 0 plus **3** is **3**.

Second, reduce each sum in the second group to the remainder when dividing by 11, like this: **2** and **8** are 10; 10 and **9** are 19; «9 minus 1 is 8» *(to find the remainder);* 8 and **7** are 15; «5 minus 1 is 4»; 4 and **5** are 9; 9 and **3** are 12; «2 minus 1 is 1»; 1 and **7** are 8; 8 and **2** are 10; 10 and **4** are 14; «4 minus 1 is 3»; and finally, 3 plus **9** is 12 whose remainder when divided by 11 is **1** («2 minus 1»).

Third *(and last),* subtract the remainder obtained in the second group (**1**) from that obtained in the first (**3**) to deduce that **93 452 678 345 678 998 821** divided by 11 produces remainder **2**.

Another way to deal with the calculation of the remainder when dividing by 11 is to take the digits from left to right and two by two to form two-digit numbers and transform them into a set of remainders that we will add and transform into remainders *a posteriori* as we operate with them, from left to right.

Let's see this with the same example:

We separate the number in terms of 2 digits from left to right (**93 45 26 78 34 56 78 99 88 21**) and subtract each ten from the units (but when these are less than those, we must add 11 to the units before operating): since 3 is less than 9, it is necessary to add 11 (3 and 11 are 14); 14 minus 9 is **5**; 5 minus 4 is **1**; 6 minus 2 is **4**; 8 minus 7 is **1**; 4 minus 3 is **1**; 6 minus 5 is **1**; 8 minus 7 is **1**; 9 minus 9 is **0**; 8 minus 8 is **0**; 1 plus 11 is 12; 12 minus 2 is **10**.

Finally, we operate with the sequence of residues obtained (**5**, **1**, **4**, **1**, **1**, **1**, **1**, **0**, **0**, **10**) adding them from left to right, and subtracting 11 when possible: **5** and **1** are 6; 6 and **4** are 10; 10 and **1** are 11 (minus 11 is 0); 0 and **1** is 1; 1 and **1** are 2; 2 and **1** are 3; 3 and **0** are 3; 3 and **0** are 3; and finally, 3 plus **10** is 13; 13 minus 11 is **2**, which is the remainder sought.

Again, all this helps us **to detect if an operation is incorrect**, but it does not guarantee the correctness of the operation, as the following example shows:

The sum of **356** plus **43** is **399**. Reducing each addend to remainders: **356** is 9 (**6** plus **3**) minus 5, namely **4**, plus 4**3** is 14 (**3** plus 11) minus 4, that is, **10**. The remainder derived from the sum of 4 and 10 is **3** (4 and 10 are 14; 4 minus 1 is 3), *the same digit* obtained from the result **399** (**9** plus **3** minus 9 is 3); this indicates that the operation could be correct (**366** ***is not correct***, and also produces remainder **3** when divided by 11). But **with a result of 369** we could say that *the **operation was NOT correct***, since the remainder when dividing **369** by 11 is **6** (**9** plus **3** minus 6), *different from* **3**.

Chapter 5
Quick division method

The objective of this chapter is to be able to deal with relatively large divisions while minimizing errors. First, we will follow the **traditional method** of division. The number to be divided is called the **dividend**; the **divisor** is the one by which we divide; the division operation tries to find how many times the dividend can contain the divisor (the **quotient** is that number) and if the dividend does not contain exactly the divisor, the **remainder** indicates the amount that is needed to achieve equality with the dividend. Basically, the operation to be performed consists of subtracting from the dividend all possible multiples of the divisor. To minimize errors, we can obtain the first precalculated multiples of the divisor (it is better to calculate ten of them, because to multiply a number by 10 we only have to add a zero, which allows checking the correctness of each of the previous sums — the multiples—). For example, let's take the number **73** as the divisor.

The first multiple is **73** (1); the others are calculated by adding 73 to the previous result, namely: 73 plus 73 is **146** (2); plus 73 is **219** (3); plus 73 is **292** (4); plus 73 is **365** (5); plus 73 is **438** (6); seven times 73 is **511** (7); plus 73 is **584** (8); plus 73 is **657** (9); and plus 73 is **730** (10) —matches the number resulting from adding a zero to its right to 73, which verifies that the calculations are correct—. *Indicating in parentheses the number of times 73 is contained provides valuable information for generating the quotient.*

Let us consider the number **402** as the **dividend**. The division is done by taking the digits from left to right. The first number (**4**) *is too small;* even taking the first two digits (**40**) the same thing happens (it does not reach 73); this forces to take three digits (**402**); now we just have to find out how many times 73 can be subtracted from that number, but it's something that we have just calculated. Searching among the precalculated multiples, *the first that does not exceed* 402 is **365**. The **5** that we put before in parentheses means **5** *times* **73**, thus indicating the sought **quotient**. Thus, **402** *divided by* **73** is **5**; but what is the **remainder**? The amount needed to complete 402 from 365, that is, **37** (402 minus 365 —a calculation that is conveniently done as follows: 400 minus 300 is 100; 100 minus 60 is 40; 40 minus 5 is 35; 35 plus 2 is **37**). In short, *73 times 5 plus 37 (the* ***remainder****)* is **402**.

Let's now consider the dividend **4 02**3 254. The first three numbers are the same as before, so we can continue from the previous calculation, but how? We must start from the last remainder, adding to its right the 3 (the digit of the dividend that follows 402), that is, **37**3. Among the precalculated multiples of 73 the one that does not exceed is 365 (5 *times* 73); the **quotient** is now **55** and the **current remainder**, 373 minus 365, that is, **8** (373 minus 360 is 13; ten minus 5 is 5 and plus 3 is **8**). The next digit to *drop* from the dividend so that it forms part of the current remainder by integrating it to its right is **2**. The **partial dividend** thus formed is **82**, from which 73 can only be subtracted once; therefore, the quotient becomes **551** and the new remainder **9** (82 minus 73; «82 minus 72 is 10, minus 1 is **9**»). We form the **partial dividend** «by dropping» the 5, namely **95**. Again, we can only subtract 73 (current quotient **5511**) **once**, generating a remainder of **22** (95 minus 73; «93 minus 73 is 20, plus 2 is 22»). Finally, we «drop» the 4 to form the partial dividend **224**. Looking at the multiples of 73 calculated above, the first of them that does not exceed 224 is 219 (**3** *times* 73). The quotient is updated to **55 113** and the **remainder** to the difference 224 minus 219, which is **5** (220 minus 210 is 10; minus 9 is 1; plus 4 is **5**). As a result of that, *73 times 55 113 plus 5 (the **remainder**)* is **4 023 254**.

Let's add two more numbers to the dividend to make it **402 325 4**65. Looking at the previous paragraph we see that the digits of the dividend that remain to be dropped are **6** and **5**; the current quotient is **55 113** and the last remainder, **5**. The first partial dividend to consider is **56** (*we have dropped* the 6). It turns out that this number is less tan 73, which is not large enough, so we have been able to subtract 73 **zero** times (this sets the current quotient to **551 130**). We dropping the next digit of the dividend (**5**), adding it to the right of the current remainder (56) to form the partial dividend **565**.

Now, consulting the precalculated multiples of 73 we see that the first one that does not exceed 565 is **511** (**7** *times* 73); the quotient is now **5 511 307** and the remainder is **54** (565 minus 511; «500 minus 500 is zero; 65 minus 11 is **54**»).

In short: *when dividing the number **402 325 465** by **73** the quotient is **5 511 307** and the remainder is **54**.*

Checking that the division is correct is easy; we just have to multiply **73** *by* **5 511 307** and *add* **54**; if the result is **402 325 465** the operation is successful. It is *more advisable* to first subtract **54** from **402 325 465** and, *using one of the multiplication methods seen above*, check that the result is equal to the product of **5 511 307** times **73**. Let's do it: **65** *minus* **54** *is* «10 plus 55 minus 54»; and as «55 minus 54 is 1»; the result of the initial difference is 10 + 1 = **11**. Thus, the operation **402 325 4**65 – 54 results in **402 325 4**11. It only remains to multiply **5 511 307** by **73**. We will use the *two-finger method* to remember some important things. This method uses a *UD* pattern (where *U* consists of taking the units digit of the product it affects and *D* taking the tens digit) that moves from right to left over the digits of the multiplicand. This pattern is associated with each digit of the multiplier (starting with the rightmost one and initially matching the *U* of the pattern with the units of the multiplicand), but for each leftmost position of the multiplier digit, the pattern shifts one position further to the right on the multiplicand. There is a *UD* pattern for each digit of the multiplier, which affects both it and the digits of the multiplicand on which to apply said pattern.

In our case, the ***UD pattern*** associated with the **3** of the multiplier starts with the ***U*** located **on the 7** of the units of the multiplicand and the *pattern* associated with the **7** of the multiplier is shifted one position to the right, so *it does not influence* the multiplicand. This determines the **units' digit of the result**, namely **1** (the units of 2**1**; the product of 3 *times* 7).

Notation: *to indicate the carry we use as many asterisk characters (*) as the amount of the carry.*

Now the patterns associated with each digit of the multiplier are shifted one position to the left. The one associated with the **3** of the multiplier at this moment places the ***U*** *on the tens' digit* of the multiplicand (**0**) and the ***D*** *on the units* of the same (**7**) at the same time that the pattern associated with the **7** of the multiplier has the ***U*** *on the digit of the units* of the multiplicand

(**7**) and the ***D*** *on nothing* (does not affect). The addends that determine the digit sought are thus: the units ***U*** of **0**0 (3 *times* 0); the tens ***D*** of **2**1 (3 *times* 7); and the units ***U*** of 4**9** (7 *times* 7); the sum of **0** plus **2** plus **9** is **11 (1** unit and **1** *carry*), defining the **tens' digit of the result** (**1***).

Again, the patterns associated with each digit of the multiplier are shifted one position to the left. The one associated with the **3** of the multiplier now has the ***U*** *on the hundreds digit* of the multiplicand (**3**) and the ***D*** *on the tens* of the same (**0**); and the pattern associated with the **7** of the multiplier is located so that the ***U*** *affects the tens' digit* of the multiplicand (**0**) and the ***D*** *affects the units* of the same (**7**). The addends that determine the digit sought are: the units ***U*** of 0**9** (3 *times* 3); the tens ***D*** of **0**0 (3 *times* 0); the units ***U*** of 0**0** (7 *times* 0) and the tens ***D*** of **4**9 (7 *times* 7); the sum **9** plus **0** plus **0** plus **4** plus **1** *carry* is **14** (**4** units and **1** *carry*, denoted **4***) and defines the **hundreds' digit of the result** (**4***). The algorithm is repeated as long as there are digits left in the multiplicand; the result confirms that **5 511 307** times **73** is **402 325 411.** All of this is summed up like this:

0	**0**	**5**	**5**	**1**	**1**	**3**	**0**	**7**	×	**7**	**3**
4	0*	2*	3*	2*	5	4*	1*	1			
								U	*D*		**21**
							U	***D***	0 7		00 + **21**
								U	*D*	**49**	
						U	***D***		3 0		09 + **00**
							U	***D***	0 7	00 + **49**	
					U	***D***			1 3		03 + **09**
						U	***D***		3 0	**21** + **00**	
				U	***D***				1 1		03 + **03**
					U	***D***			1 3	07 + **21**	
			U	***D***					5 1		**15** + **03**
				U	***D***				1 1	07 + **07**	
		U	***D***						5 5		**15** + **15**
			U	***D***					5 1	35 + **07**	
	U	***D***							0 5		00 + **15**
		U	***D***						5 5	35 + **35**	
	U	***D***							0 5	00 + **35**	

To facilitate reading, in the previous table the digits of the multiplicand that come into play in each calculation have been indicated under the column of the symbol « × ».

Additionally, *one of the error detection methods* seen above can be used, converting each number to a single digit (dividend, divisor, quotient and remainder) to verify that all these digits comply with the classic rule: *dividend is equal to the divisor by the quotient plus the remainder.*

Quick division method

Just as we saw the two-finger method for multiplying without a calculator, there is a similar method for dividing from memory; but this time we need to augment the pattern notation, adding the ability to catch all ***all*** (***T***) digits of the number and **not just** the ***units*** (***U***) or ***tens*** (***D***). The pattern must be applied to the product of the digit of the divisor affected by the last digit of the quotient being calculated. In the resolution of a division, two patterns will be used: *a unique one* (***TD***) associated so that ***T*** *coincides with the digit with the highest weight of the divisor* (the one on the left) and *several* ***UD*** (one for each remaining digit of the divisor so that the position of ***U*** in the pattern *is the same as* that of said digit) which are located from left to right *(the first of them coinciding with the position* ***D*** *of the* ***TD*** *pattern).*

Single digit divisor

Only the ***TD*** pattern (actually just ***T***) is applicable here, so it **matches the traditional division method**. Let's consider that the *dividend* is **67** and the *divisor* is **5**:

To find the **first digit of the quotient**, we divide the leftmost digit(s) of the dividend by the divisor (**6** *divided by* **5** is **1** — not 2, because 5 *times* 2 is 10, exceeding 6).

Now we **subtract** «the result of applying the ***T*** *pattern* to the product **5** *times* **1**» (***T*** selects **all the digits** in this calculation, that is, **5**) « from the last partial dividend» (**6** minus **5** is **1**) and we form a **new partial dividend** with this number added to the next digit of the dividend, that is, **17**.

We look for how many times this number contains the divisor (four times 5 is 20, exceeding 17, so **3** is the *desired quotient digit* —now **13** is the total «partial» quotient). Applying the ***T pattern*** on (5 *times* 3) «the product of the *divisor* by the *last digit of the* found *quotient*» is to take the whole number, namely **15**, which subtracted from the partial dividend (**17**) gives **2**.

Thus, the result of the division is **13** for the quotient and **2** for the remainder, which is correct, since 13 *times* 5 is «**50** (10 *times* 5) plus **15** (3 *times* 5)», that is, **65**; plus **2** is **67** (the dividend).

Two-digit divisor

In this case, two patterns are applied: ***TD*** coinciding ***T*** *on the tens and* ***D*** *on the units of the divisor;* ***UD*** with ***U*** *on the units of the divider and* ***D*** *in the void.*

Let's consider the last division example seen above (dividend **402 325 465** and divisor **73**):

For the **first digit** of the quotient, we need to find the number of times (not 0) that the tens' digit of the divisor (**7**) is contained in the first leftmost digits of the dividend. Since 4 is less than 7 we must also consider the next digit and see how many times the number thus formed (**40**) contains 7 (6 *times* 7 is 42, which exceeds 40). The **first digit of the quotient** is therefore **5**.

4	**0**	**2**	**3**	**2**	**5**	**4**	**6**	**5**	÷	**7**	**3**		
		5	5	1	1	3	0	7	*Quotient*	***T***	***D***	***U***	*Step*
	4	2							5	**35**	**15**		1
	3	**7**										**15**	2
		1	3						5	**35**	**15**		1
			8									**15**	2
			1	2					1	**07**	**03**		1
				9								03	2
				2	5				1	**07**	**03**		1
				2	**2**							03	2
					1	4			3	**21**	**09**		1
						5						09	2
						5	6		0	**00**	**00**		1
						5	**6**					00	2
							5	5	7	**49**	**21**		1
							5	**4**				21	2

Quick division method

Trachtenberg

The ***TD*** *pattern* is such that ***T*** *(take all digits) applies to the tens* and ***D*** *(take only the tens) applies to the units of the divisor; and the* ***U*** *pattern (take only the units) on the units* of the divisor. These patterns ***are applied in 2 steps*** as follows:

The *FIRST step* consists of **multiplying the found quotient** (5) **by each digit of the divisor** (73) and **after applying the** corresponding **pattern** —***T*** over the product by the tens (5 *times* 7 *is* **35**) is **35** and ***D*** over the product by the units (5 *times* 3 *is* **1**5) is the tens' digit, that is, **1**—, **add both** (35 plus 1 is **36**). This sum **must be subtracted from the last partial dividend** (40 *minus* 36 *is* **4**); the result **is added to the left of the next digit of the dividend** (**4**2).

In the *SECOND step* it is enough **to multiply the found quotient** (5) **by the units of the divisor** (3) and **after applying the** corresponding **pattern** (***U*** over 5 *times* 3 are the units of 1**5**, that is, **5**) **subtract from the number found in the first step** (42 *minus* 5 *is* **37**). The result is the leftmost part **of the new partial dividend** *with which* ***to repeat the entire division process*** (it is really ***the remainder*** *of the division* at this point, since 402 is 73 *times* 5 plus 37).

The digits of the dividend already used are **402**, the rest are **325 465**; the **last partial dividend** is the last remainder (**37**); the **current quotient** is **5** and the **divisor** is the same (**73**).

For the **second digit** of the quotient, we must look for the number of times (not 0) that the tens' digit of the divisor (**7**) is contained in the first leftmost digits of the last partial dividend (**37**). Since **3** is less than 7 (of the divisor **7**3) we must also consider the next digit **7** (de 3**7**) and see how many times the number thus formed (**37**) contains 7 (6 *times* 7 is 42; it exceeds 37). Thus, the **second digit of the quotient** is **5**.

In the *FIRST step*, **we multiply the found quotient** (5) **by each digit of the divisor** (73) and **after applying the** corresponding **pattern** —***T*** over the product by the tens (5 *times* 7 *is* **35**) is **35** and ***D*** over the product by the units (5 *times* 3 *is* **1**5) is the tens' digit, that is, **1**—, **we add both** (35 plus 1 is **36**). This sum **is subtracted from the last partial dividend** (37 *minus* 36 *is* **1**) and the result **is added to the left of the next digit of the dividend** (**1**3).

In the *SECOND step* it is enough **to multiply the found quotient** 5 **by the units of the divisor** (3) and

after applying the corresponding **pattern** (***U*** over 5 *times* 3 are the units of 15, that is, **5**) **subtract from the number found in the first step** (13 *minus* 5 *is* **8**). This result is the leftmost part **of the new partial dividend** *with which* ***the entire division process must again be repeated*** (now **8** is ***the remainder*** *of the division* at this point, since 4023 is 73 *times* 55 plus 8).

The digits of the dividend already used are **4023**, the rest are **25 465**; the **last partial dividend** *is the last remainder* (**8**); the **current quotient** is **55** and the **divisor** is the same (**73**).

For the **third digit** of the quotient, we need to find the number of times (not 0) that the tens' digit of the divisor (**7**) is contained in the first leftmost digits of the last partial dividend (**8**). The 8 can contain 7 only **once**, so the **third digit of the quotient** is **1**.

In the *FIRST step*, **we multiply the found quotient** (1) **by each digit of the divisor** (73) and **after applying the** corresponding **pattern** —***T*** over the product by the tens (1 *times* 7 *is* **07**) is **07** and ***D*** over the product by the units (1 *times* 3 *is* **0**3) is the tens' digit, that is, **0**—, **we add both** (7 plus 0 is **7**). This sum **is subtracted from the last partial dividend**

(8 *minus* 7 *is* **1**) and the result **is added to the left of the next digit of the dividend** (**1**2).

In the *SECOND step* it is enough **to multiply the found quotient** (1) **by the units of the divisor** (3) and **after applying the** corresponding **pattern** (***U*** over 1 *times* 3 are the units of 0**3**, that is, **3**) **subtract from the number found in the first step** (12 *minus* 3 *is* **9**). This result is the leftmost part **of the new partial dividend** *with which **the entire division process must** again **be repeated*** (now **9** is ***the remainder** of the division* at this point, since 40 232 is 73 *times* 551 plus 9; the unused digits of the dividend are **5465**).

The rest of the calculation is similar *(see the example division summary table above):*

The **fourth digit** of the quotient (**9** *divided by* 7 is 1, since 9 contains 7 only once) is **1**; **7** (***T*** over 1 *times* 7) plus **0** (***D*** over 1 *times* 3) is **7**; **9** *minus* 7 is **2**; **2**5 (which comes from adding the previous **2** to the left of the next digit of the dividend) minus **3** (***U*** over 1 *times* 3) is **22** (the remainder; the *unused digits of the dividend* are **465**).

The **fifth digit** of the quotient (**22** *divided by* 7) is **3**; **21** (***T*** over 3 *times* 7) plus **0** (***D*** over 3 *times* 3 are the tens of **03**) is **21**; **22** *minus* 21 is **1**; **14** (which comes from adding the previous **1** to the left of the next digit of the dividend) minus **9** (***U*** over 3 *times* 3) is **5** (the remainder; *the unused digits* of the dividend are **65**).

The **sixth digit** of the quotient is **0** (**5** is less than 7); **0** (***T*** over 0 *times* 7) plus **0** (***D*** over 0 *times* 3) is **0**; **5** *minus* 0 is **5**; **5**6 (the previous **5** added to the left of the next digit of the dividend) minus **0** (***U*** over 0 *times* 3) is **56** (the remainder; *the digit* of the dividend *not yet used* is **5**).

The **seventh digit** of the quotient is **7** (**56** *divided by* 7); **49** (***T*** over 7 *times* 7) plus **2** (***D*** over 7 *times* 3) is **51**; **56** *minus* 51 is **5**; **5**5 (the newly calculated **5** added to the left of the next digit of the dividend) minus **1** (***U*** over 7 *times* 3) is **54** (the **final remainder**).

*The **seventh digit** of the quotient **cannot be 8** because 56 (**T** over 8 times 7) plus **2** (**D** over 8 times 3) equals **57**; and we can't subtract 57 from 56 (it would be negative).*

*« In short, **402 325 411** divided by **73** is **5 511 307**»*

Three or more digits divisor

In this case **two patterns** will be used: *a single one* (***TD***) associated so that ***T*** matches *the leftmost digit* of the divisor, which **is applied in the first step** and *several* ***UD*** (one for each remaining digit of the divisor so that the ***U*** *position of the pattern is the same* as that of said digit), which are located from left to right *(the first of them coinciding with the **D** position of the **TD** pattern)* and **are applied in the subsequent steps**; *for example*, with a *five-digit* **divisor**, the **pattern distribution** would look like this:

	TT	*UT*	*Hundreds*	*Tens*	*Units*
Step **1**	***T***	***D***			
Step **2**		***U***	***D***		
Step **3**			***U***	***D***	
Step **4**				***U***	***D***
Step **5**					***U***

T *tens and units*; ***D*** *only tens*; ***U*** *only units*

The pattern **acts on the product** of *the digit of the corresponding by the last calculated digit-quotient.*

3	8	6	7	3	0	3	2	÷	4	1 7	5	*Divisor's*
				9	2	6	3	*Quotient*	***T***	***U***	***D***	*digits*
	2	6							**36**		**09**	4 1
	1	**1**	7					9		0**9**	**63**	1 7
	1	**1**	**0**	3						6**3**	**4**5	7 5
	1	**0**	**9**	**8**						4**5**		5
		2	9						**08**		**02**	4 1
		2	**6**	8				2		0**2**	**14**	1 7
		2	**6**	**3**	0					1**4**	**1**0	7 5
		2	**6**	**3**	**0**					1**0**		5
			2	3					**24**		**06**	4 1
			1	**3**	0			6		0**6**	**42**	1 7
			1	**2**	**5**	3				4**2**	**3**0	7 5
			1	**2**	**5**	**3**				3**0**		5
				0	5				**12**		**03**	4 1
					0	3		3		0**3**	**21**	1 7
						1	2			2**1**	**1**5	7 5
							7			1**5**		5

The previous table shows the process of dividing by a four-digit number, an example that we are going to observe in detail to clarify the algorithm. The dividend is **38 673 032** and the divisor is **4175**.

For the **first digit** of the quotient, we must look for the number of times (not 0) that the most significant digit divisor (**4**) is contained in the first leftmost digits of the dividend. Since 3 is less than 4 we must also consider the next digit and see how many times the number thus formed (**38**) contains 4 (9 *times* 4 is 36 — less than 38). The **first digit of the quotient** is thus **9**.

Each *calculated* ***digit of the quotient marks a cycle*** *of as many steps as there are digits in the divisor:*

In the *FIRST step,* **we multiply the found quotient** (9) **by each of the two leftmost digits of the divisor** (41) and **after applying the** corresponding **pattern** (***T*** to the product by the ***leftmost digit*** *of the selected divisor* —9 *times* 4 is **36**— and ***D*** to *the product by the* ***digit to the right*** *of the selected divisor* —9 *times* 1 is **0**9; the tens' digit is **0**), **we add both** (36 *plus* 0 is **36**). This sum **is subtracted from the last partial dividend** (38 *minus* 36 is **2**) and the result **is added to the left of the next digit of the dividend**

(**2**6).

In the *SECOND step*, **we multiply the quotient found** (9) **by each of the two digits of the divisor chosen** *from the second* **to the left** (17) and **after applying the** corresponding **pattern** (***U*** to the product by ***the digit on the left*** *of the selected divisor* — 9 *times* 1 is 0**9**; the units digit is **9**— and ***D*** to *the product by* ***the right digit*** *of the selected divisor* — 9 *times* 7 is **6**3; the tens' digit is **6**), **the two are added** (9 *plus* 6 is **15**) and the result **is subtracted from the last partial dividend** (26 *minus* 15 is **11**) and **attached to the left of the next digit of the dividend**, that is, **11**7.

In the *THIRD step*, **we multiply the quotient found** (9) **by each of the two digits of the divisor chosen** *starting from the third one* **to the left** (75) and **after applying the** corresponding **pattern** (***U*** to *the product by* ***the digit to the left*** *of the selected divisor* —9 *times* 7 is 6**3**; the units' digit is **3**— and ***D*** to the product by ***the digit to the right*** *of the selected divisor* —9 *times* 5 is **4**5; the tens' digit is **4**), **the two are added** (3 *plus* 4 is **7**) and the result **is subtracted from the last partial dividend** (117 *minus* 7 is **110**) and

attached to the left of the next digit of the dividend (**110**3).

In the *FOURTH step*, since the divisor has four figures, it is enough **to multiply the quotient found** (9) **by the units of the divisor** (5) and **after applying the** corresponding **pattern** (***U*** over 9 *times* 5; the units of 4**5** are **5**) **subtract of the number found in the previous step** (1103 *minus* 5 is **1098** —100 *minus* 5 is 95, plus 3 is 98 and plus 1000 is **1098**). The result is the leftmost part of **the new partial dividend** with which ***the entire division process must be repeated*** (now **1098** is ***the remainder*** *of the division (at this point),* since 38 673 is 37 575 (9 *times* 4175) plus 1098; the *not yet used digits* of the main dividend are **032**).

« The current dividend is **1098 032**.»

For the **second digit** of the quotient, we must look for the number of times (not 0) that the most significant digit of the divisor (**4**) is contained in the first leftmost digits of the current dividend **1098**. Since 1 is less than 4 we must also consider the next digit and see how many times the number thus formed (**10**) contains 4 (3 *times* 4 exceeds 10). Therefore, the **second digit of the quotient** is **2**.

Again, the 4 previous steps are repeated (as many as the numbers in the divisor):

In the *FIRST step*, **we multiply the found quotient** (2) **by each of the two leftmost digits of the divisor** (41) and **after applying the** corresponding **pattern** (***T*** to the product by the ***leftmost digit*** *of the selected divisor* —2 *times* 4 are **08**— and ***D*** to *the product by the* ***digit to the right*** *of the selected divisor* —2 *times* 1 is **0**2; the tens' digit is **0**), **we add both** (8 *plus* 0 is **8**). This sum **is subtracted from the last partial dividend** (10 *minus* 8 is **2**) and the result **is attached to the left of the next digit of the dividend** (**2**9).

In the *SECOND step*, **we multiply the quotient found** (2) **by each of the two digits of the divisor chosen** *starting from the second one* **on the left** (17) and **after applying the** corresponding **pattern** (***U*** to *the product by* ***the digit to the left*** *of the selected divisor* —2 *times* 1 is 0**2**; the units' digit is **2**— and ***D*** to *the product by* ***the digit to the right*** *of the selected divisor* —2 *times* 7 is **1**4; the tens' digit is **1**), **the two are added** (2 *plus* 1 equals **3**) and the result **is subtracted from the last partial dividend** (29 *minus* 3 equals **26**) and **added to the left of the next digit of the dividend**, that is, **26**8.

In the *THIRD step*, **we multiply the found quotient** (2) **by each of the two digits of the divisor chosen** *starting from the third one* **on the left** (75) and **after applying the** corresponding **pattern** (***U*** to *the product by* ***the digit on the left*** *of the selected divisor* —2 *times* 7 is 1**4**; the units' digit is **4**— and ***D*** to *the product by* ***the digit to the right*** *of the selected divisor* —2 *times* 5 is **1**0; the tens' digit is **1**), **both are added** (4 *plus* 1 equals **5**) and the result **is subtracted from the last partial dividend** (268 *minus* 5 equals **263**) and **added to the left of the next digit of the dividend** (**263**0).

In the *FOURTH step*, since the divisor has four figures, it is enough **to multiply the found quotient** (2) **by the units of the divisor** (5) and **after applying the** corresponding **pattern** (***U*** over 2 *times* 5; the units of 1**0** are **0**) **subtract from the number found in the previous step** (2630 *minus* 0 equals **2630**). The result is the leftmost part of the **new partial dividend** with which ***the entire division process must be repeated*** (now **2630** is ***the remainder*** *of the division (at this point),* since 386 730 is 384100 (92 *times* 4175) plus 2630; the *not yet used digits* of the main dividend are **32**).

« The current dividend is **2630 32**.»

For the **third digit** of the quotient, we must look for the number of times (not 0) that the most significant digit of the divisor (**4**) is contained in the first leftmost digits of the current dividend (**2630**). Since 2 is less than 4, we must also consider the next digit and see how many times the number thus formed (**26**) contains 4 (7 *times* 4 exceeds 26). This is why **second digit of the quotient** is **6**.

Again the 4 previous steps are repeated (as many as the digits in the divisor):

In the *FIRST step*, **we multiply the found quotient** (6) **by each of the two leftmost digits of the divisor** (41) and **after applying the** corresponding **pattern** (***T*** to *the product by* ***the leftmost digit*** *of the selected divisor*—6 *times* 4 is **24**— and ***D*** to *the product by* ***the digit to the right*** *of the selected divisor*— 6 *times* 1 is **0**6; the tens' digit is **0**), **we add both** (24 *plus* 0 is **24**). This sum **is subtracted from the last partial dividend** (26 *minus* 24 is **2**) and the result **is attached to the left of the next digit of the dividend**

(**23**).

In the *SECOND step*, **we multiply the found quotient** (6) **by each of the two digits of the divisor chosen** *starting from the second* **on the left** (17) and **after applying the** corresponding **pattern** (***U*** to *the product by* ***the digit on the left*** *of the selected divisor* —6 *times* 1 is 0**6**; the units' digit is **6**— and ***D*** to *the product by* ***the digit to the right*** *of the selected divisor* —6 *times* 7 is **4**2; the tens' digit is **4**), **both are added** (6 *plus* 4 are **10**) and the result **is subtracted from the last partial dividend** (23 *minus* 10 is **13**) and **attached to the left of the next digit of the dividend**, that is, **13**0.

In the *THIRD step*, **we multiply the found quotient** (6) **by each of the two digits of the divisor chosen** *starting from the third* **on the left** (75) and **after applying the** corresponding **pattern** (***U*** to *the product by* ***the digit on the left*** *of the selected divisor* — 6 *times* 7 is 4**2**; the units' digit is **2**— and ***D*** to *the product by* ***the digit to the right*** *of the selected divisor* —6 *times* 5 is **3**0; the tens' digit is **3**), **both are added** (2 *plus* 3 are **5**) and the result **is subtracted from the last partial dividend** (130 *minus* 5 is **125**) and **attached**

to the left of the next digit of the dividend (**12**53).

In the *FOURTH step*, since the divisor has four figures, it is enough **to multiply the quotient found** (6) **by the units of the divisor** (5) and **after applying the** corresponding **pattern** (***U*** over 6 *times* 5; the units of 3**0** are **0**), **subtract from the number found in the previous step** (1253 *minus* 0 is **1253**). The result is the leftmost part of **the new partial dividend** with which ***to repeat the entire division process*** (now **1253** is ***the remainder*** *of the division (at this point),* since 3 867 303 is 3 866 050 (926 *times* 4175) plus 1253; the *not yet used digit* of the main dividend is **2**).

« The current dividend is **1253 2**.»

For the **fourth digit** of the quotient (the last integer) we must look for the number of times (not 0) that the most significant digit of the divisor (**4**) is contained in the first leftmost digits of the current dividend (**1253**). Since 1 is less than 4 we must also consider the next digit and determine how many times the number thus formed (**12**) contains 4 (3 *times* 4 is 12). The **second digit of the quotient** is thus **3**.

To finish, it only remains to repeat the four previous steps (as many as there are figures in the divisor):

In the *FIRST step,* **we multiply the found quotient** (3) **by each of the two leftmost digits of the divisor** (41) and **after applying the** corresponding **pattern** (***T*** to the product by ***the leftmost digit*** *of the selected divisor* —3 *times* 4 are **12**— and ***D*** to *the product by* ***the digit to the right*** *of the selected divisor* —3 *times* 1 is **0**3; the tens' digit is **0**), **we add both** (12 *plus* 0 is **12**). This sum **is subtracted from the last partial dividend** (12 *minus* 12 is **0**) and the result **is attached to the left of the next digit of the dividend** (**0**5).

In the *SECOND step,* **we multiply the found quotient** (3) **by each of the two digits of the divisor chosen** *starting from the second* **on the left** (17) and **after applying the** corresponding **pattern** (***U*** the product by ***the digit on the left*** *of the selected divisor* —3 *times* 1 is 0**3**; the units' digit is **3**— and ***D*** to *the product by* ***the digit to the right*** *of the selected divisor* —3 *times* 7 is **2**1; the tens' digit is **2**), **both are added** (3 *plus* 2 equals **5**) and the result **is subtracted from the**

last partial dividend (5 *minus* 5 is **0**) and **attached to the left of the next digit of the dividend,** that is, **0**3.

In the *THIRD step,* **we multiply the found quotient** (3) **by each of the two digits of the divisor chosen** *starting from the third* **on the left** (75) and **after applying the** corresponding **pattern** (***U*** to *the product by* ***the digit on the left*** *of the selected divisor* — 3 *times* 7 is 2**1**; the units' digit is **1**— and ***D*** to *the product by* ***the digit to the right*** *of the selected divisor* —3 *times* 5 is **1**5; the tens' digit is **1**), **both are added** (1 *plus* 1 is **2**), **is subtracted from the last partial dividend** (3 *minus* 2 is **1**) and the result **is attached to the left of the next digit of the dividend** (**1**2).

In the *FOURTH step,* since the divisor has four figures, it is enough **to multiply the found quotient** (3) **by the units of the divisor** (5) and **after applying the** corresponding **pattern** (***U*** over 3 *times* 5 is **5**) **subtract from the number found in the previous step** (12 *minus* 5 is **7**). The result is the leftmost part of **the new partial dividend** with which ***the entire division process would have to be repeated*** to obtain the decimals of the division, but this is not the case (**7** is **the remainder** of the division).

0	0	0	0	4	1	7	5	×	9 2	6 3
3	8	6**	7*	3**	0*	2	5			
							U	*D*		15
						U	*D*	7 5 × 3		21 + 15
							U	5 × 6		30
					U	*D*		1 7 × 3		03 + 21
						U	*D*	7 5 × 6		42 + 30
				U	*D*			4 1 × 3		12 + 03
					U	*D*		1 7 × 6		06 + 42
						U	*D*	7 5 × 2	14 + 10	
							U	5 × 9	45	
			U	*D*				0 4 × 3		00 + 12
				U	*D*			4 1 × 6		24 + 06
					U	*D*		1 7 × 2	02 + 14	
						U	*D*	7 5 × 9	63 + 45	
			U	*D*				0 4 × 6		00 + 24
				U	*D*			4 1 × 2	08 + 02	
					U	*D*		1 7 × 9	09 + 63	
				U	*D*			4 1 × 9	36 + 09	

The table above shows the product of the divisor (**4175**) by the quotient (**9263**). The result provided by this calculation is **38 673 025**, which added to the rest of the division (**7**) is **38 673 032** (the dividend). This ensures that all the operations carried out are correct.

Chapter 6

Squares and its roots

If we represent a number N by a straight line of equal length to its magnitude, *we can put together N of those lines* (each one adjacent to the previous one) to form a surface. Its area is N *times* N and it turns out to be a square. The product «**N** times **N**» is called ***square*** *of a number.*

The *inverse operation* of the square of a number is called the ***square root*** of said number and consists of *finding the number N which when squared results in N times N.*

To *fix concepts*, nothing better than an **example**: *the product* **7** times **7** *is called* ***the square of 7*** *and is* ***49****; the number* **7** ***is its square root***.

The square of a number

One-digit numbers

The square of a one-digit number is simple. *There are **only** 10 one-digit numbers* (**0**, **1**, **2**, **3**, **4**, **5**, **6**, **7**, **8** and **9**) and its respective squares are **0**, **1**, **4**, **9**, **16**, **25**, **36**, **49**, **64** and **81**.

Two-digit numbers

Let us first recall how to quickly multiply two two-digit numbers with an example, namely, **32** *times* **19** (whose product is **608**):

The **units of the result** (**8**) are the product of the units of each one of the factors (**2** of 3**2** and **9** of 1**9**); **2** *times* **9** is **18** (**1** ***carry*** occurs).

The **tens of the result** (**0**) are the sum of the previous carry (**1**) and two more addends (generated by multiplying the tens of one of the factors by the units of the other), namely, **27** (**3** *times* **9** —the **extreme digits of** **3**2 1**9**); and **2** (**2** *times* **1** —the **middle digits of** 3**2** **1**9); **27** plus **2** plus **1** *carry* equals **30** (**0** units and **3** ***carry***).

The **hundreds of the result** (**6**) are the sum of

the previous carry (**3**) with the product of the tens' digits of the operands (3 *times* 1 is **3**); 3 plus 3 is **6**.

If *multiplicand and multiplier* **are** *two* **identical** *two-digit numbers,* we are calculating the **square** of a *two-digit* **number**:

- The **units of the result** are the *square of its units; carry can be generated.*
- The **tens in the result** are the *previous carry* added to ***twice*** *"the product of its units times its tens"; carry can be generated.*
- The **hundreds and thousands of the result** are the *square of its tens* plus *the* previous *carry* (if any).

Two-digit numbers ending in 5

To calculate the square of a number ending in 5 it is enough to multiply its tens' digit by the number resulting from "*adding* **1**" to said digit and to the right of this result attach **25** as *tens and units*. For example: **1**5 *times* **1**5 is **2**25 (because 1 *times* "1 + **1**" is **2**) and **7**5 *times* **7**5 is **56**25 (since 7 *times* "7 + **1**" is **56**).

Let's see *why* this rule *works:*

The **units of the result** will **always** be **5**, since

they come from the product of 5 *times* 5, that is 2**5** (**5** units and 2 *carry*).

The **tens of the result** are **always 2**, because as we saw, they are 2 times "5 times the ***tens' digit*** *of the number we are squaring"*, a digit that ***determines the current carry***, since 2 *times* 5 equals 10 and multiplying a number for 10 is to add a 0 to its right, which is why the *tens of the result* are always the previous carry, that is, **2** and the ***current carry are the tens*** *of the original number.*

the **hundreds and thousands of the result** are the *tens' digit* of the number we are squaring, *multiplied by itself, and added to itself (the **previous carry**).* But denoting D to the tens, this is (D *times* D) *plus* D, which, taking a common factor to D is D *times* (D *plus* 1), that is, *the tens' digit multiplied by the number resulting from adding 1 to said digit.*

With a *concrete **example*** it looks much better: **35** *times* **35** is **12**25, since (3 *times* 3) *plus* 3 —or also equivalently 3 *times* (3 *plus* 1)— is **12**.

Two-digit numbers whose ten is 5

To calculate the square of a number whose ten is 5, it is enough to put as *units and tens* of the result, the **square of its units** and as *hundreds and thousands* of the result, said **units** plus **25**.

Assuming that *the number we want to square is N*, let's see *why this* rule *works:*

The **units of the result** will **always** be *the units of the product of the digit of the units of N by itself.*

The **tens of the result** are calculated by adding to the previous carry the product of the units of N times 5 and times 2 (that is, by 10); since multiplying a number by 10 to attach a 0 to its right, the **tens of the result** are the previous carry (which comes from the tens of "the square of the digit of the units of N" —*even if they are zero*), and **the current carry** is the product of the **units of N** by the 1 of **10**.

The **hundreds and thousands of the result** are calculated **by adding 25** (the tens' digit of N, *squared*) **to the units of N** (the previous carry).

Quick examples *(all cases):*

(**50** *times* **50** *is* **2500**): 25 *plus* 0 is **25**; 0 *times* 0 is **00**; the *square of* **50** is **25**00.

(**51** *times* **51** *is* **2601**): 25 *plus* 1 is **26**; 1 *times* 1 is **01**; the *square of* **51** is **26**01.

(**52** *times* **52** *is* **2704**): 25 *plus* 2 is **27**; 2 *times* 2 is **04**; the *square of* **52** is **27**04.

(**53** *times* **53** *is* **2809**): 25 *plus* 3 is **28**; 3 *times* 3 is **09**; the *square of* **53** is **28**09.

(**54** *times* **54** *is* **2916**): 25 *plus* 4 is **29**; 4 *times* 4 is **16**; the *square of* **54** is **29**16.

(**55** *times* **55** *is* **3025**): 25 *plus* 5 is **30**; 5 *times* 5 is **25**; the *square of* **55** is **30**25.

(**56** *times* **56** *is* **3136**): 25 *plus* 6 is **31**; 6 *times* 6 is **36**; the *square of* **56** is **31**36.

(**57** *times* **57** *is* **3249**): 25 *plus* 7 is **32**; 7 *times* 7 is **49**; the *square of* **57** is **32**49.

(**58** *times* **58** *is* **3364**): 25 *plus* 8 is **33**; 8 *times* 8 is **64**; the *square of* **58** is **33**64.

(**59** *times* **59** *is* **3481**): 25 *plus* 9 is **34**; 9 *times* 9 is **81**; the *square of* **59** is **34**81.

The following table **contains the squares** of the first hundred integers (from **0** to **99**); learning it considerably increases the speed when doing memory calculations; **the rows indicate the tens** of the number and **the columns its units**, that is, *the row box to the right of the 3 and the column below the 8 locates the square of 38, namely 1444.*

Square** of the numbers from **0** to **99

	0	**1**	**2**	**3**	**4**	**5**	**6**	**7**	**8**	**9**
0	0	1	4	9	16	**25**	36	49	64	81
1	100	121	144	169	196	**225**	256	289	324	361
2	400	441	484	529	576	**625**	676	729	784	841
3	900	961	1024	1089	1156	**1225**	1296	1369	1444	1521
4	1600	1681	1764	1849	1936	**2025**	2116	2209	2304	2401
5	**2500**	**2601**	**2704**	**2809**	**2916**	**3025**	**3136**	**3249**	**3364**	**3481**
6	3600	3721	3844	3969	4096	**4225**	4356	4489	4624	4761
7	4900	5041	5184	5329	5476	**5625**	5776	5929	6084	6241
8	6400	6561	6724	6889	7056	**7225**	7396	7569	7744	7921
9	8100	8281	8464	8649	8836	**9025**	9216	9409	9604	9801

*The cell (**row D, column U**) is the square* $(\mathbf{10} \cdot \boldsymbol{D} + \boldsymbol{U})^2$

$\left(\boldsymbol{E.g.}, \textit{the cell } (\mathbf{3}, \mathbf{1}) \textit{ locates } \mathbf{31^2} = 961\right)$

Numbers with three or more digits

The algorithm that we are going to explore here is based on the method that we have just seen to obtain the square of a two-digit number. We chose **265** as the three-digit victim because it is quite calculation-friendly.

Step	Calculations		Trachtenberg				
1	**265**	65^2		**4**	**2**	**2**	**5**
2	**265**	$2 \cdot (2 \cdot 5)$		2	0		
		+		**6**	**2**	**2**	**5**
3	**265**	$2 \cdot (2 \cdot 6)$	2	4			
		+	**3**	**0**	**2**	**2**	**5**
4	265	2^2	4	+			
	265	265^2	**7**	**0**	**2**	**2**	**5**

The **first step** consists of calculating the square of the two rightmost digits of the number (**65**). Since it ends in 5, we only need to multiply **6** *by* 7 and attach **25** to get the result, namely **4225**.

In the **second step** the product of the *hundreds'* digit (**2**) by the *units'* digit (**5**) of **265** (called **the cross product**) is calculated and **doubled** (**2** *times* **5** is **10**; *twice* **10** equals **20**). This number is added to the most significant digits of **42**25 (**42** plus **20** equals **62**) keeping the remainder, that is, **62**25.

*From here on, the algorithm is used to obtain the square of a two-digit number over the two most significant digits of **265**, but ignoring the units squared, that is, the calculation is reduced to doubling the product **2** times **6** and square the tens of **26**; let's see how to do this:*

In the **third step** the product of the *hundreds'* digit (**2**) by the *tens'* digit (**6**) of **26**5 (**2** *times* **6** equals **12**) is calculated and **doubled** (*double* **12** equals **24**); this result is added to **6**225 from the left so that the **4** of 2**4** *coincides in position with the most significant digit* of **6**225 (**24** *plus* **6** equals **30**), keeping the rest, that is **30** 225.

In the **fourth step** *the **square** of the hundreds' digit* of **2**65 (**2** *times* **2** is **4**) *is calculated and added* to the most significant digit of **3**0 225 (**3** *plus* **4** equals **7**), keeping the remainder, namely, **70 225**.

As a check, let's multiply **265** by itself using the *two-finger method* to see that the product is equal to **70 225**:

0	**0**	**0**	**2**	**6**	**5**	×	**2**	**6**	**5**
0	7	0*	2*	2	5				
					U	*D*			25
				U	*D*	6 5 × 5			30 + 25
					U	5 × 6		30	
			U	*D*		2 6 × 5			10 + 30
				U	*D*	6 5 × 6		36 + 30	
					U	5 × 2	10		
		U	*D*			0 2 × 5			00 + 10
			U	*D*		2 6 × 6		12 + 36	
				U	*D*	6 5 × 2	12 + 10		
		U	*D*			0 2 × 6		00 + 12	
			U	*D*		2 6 × 2	04 + 12		
		U	*D*			0 2 × 2	00 + 04		

Two-finger method **Trachtenberg**

The algorithm that we have seen to find the square of a three-digit number **can be extended to four or more digits**:

Step	*Calculations*		*Trachtenberg*							
1	3265	65^2					**4**	**2**	**2**	**5**
Step 2	3265	$2\cdot(2\cdot5)$					2	0		
		+					**6**	**2**	**2**	**5**
Step 3	3265	$2\cdot(2\cdot6)$				2	4			
		+				**3**	**0**	**2**	**2**	**5**
Step 4	3265	2^2				4				
	3265	265^2				**7**	**0**	**2**	**2**	**5**
Step 5	3265	$2\cdot(3\cdot65)$			3	9	0			
		+			**4**	**6**	**0**	**2**	**2**	**5**
Step 6	3265	$2\cdot(3\cdot2)$		1	2					
		+		**1**	**6**	**6**	**0**	**2**	**2**	**5**
Step 7	3265	3^2	0	9						
	3265	3265^2	**1**	**0**	**6**	**6**	**0**	**2**	**2**	**5**

We can take advantage of the first four steps. The new digit (**3**) generates *an addend for each step:*

- In *step* **5**, we multiply it by each least significant digit of 2**65** and double (see above).
- In *steps* **6** and **7**, we join it to the most significant digit of **2**65 forming the number 32 which is **squared** (ignoring the units' square).

We can take advantage of the previous calculations to show that *the algorithm is valid for one more number:*

Steps **1** *to* **7** *(previous)*	*Calculations*			*Trachtenberg*							
	3265^2			**1**	**0**	**6**	**6**	**0**	**2**	**2**	**5**
Step **5** *(new)*	2 · (**5** · **265**)			2	6	5	0				
	+			**3**	**7**	**1**	**6**	**0**	**2**	**2**	**5**
Step **6** *(new)*	2 · (**5** · **3**)		3	0							
	+		**3**	**3**	**7**	**1**	**6**	**0**	**2**	**2**	**5**
Step **7** *(new)*	**5** · **5**	2	5	+							
	53265^2	**2**	**8**	**3**	**7**	**1**	**6**	**0**	**2**	**2**	**5**

Square root of a number

If we consider the number $\boldsymbol{N}$ and multiply it by itself we get $\boldsymbol{N} \cdot \boldsymbol{N}$ *(that is, we have squared it).* The **square root** of the **radicand** $\boldsymbol{N} \cdot \boldsymbol{N}$ is $\boldsymbol{N}$, which is denoted as: $\sqrt{\boldsymbol{N} \cdot \boldsymbol{N}} = \sqrt{\boldsymbol{N}^2} = \sqrt[2]{\boldsymbol{N}^2} = \boldsymbol{N}^{\frac{2}{2}} = \boldsymbol{N}^1 = \boldsymbol{N}$.

$\sqrt{2}$	6	1	8	0	3	3	Calculations	**1 6 1 8**	
1							1^2	*Partial* ***root***	
1	6	1					$R = 2 - 1$	(· **2**)	**Raíz**
1	**5**	**6**					$26 \cdot 6$	**2**	1
		5	8	0			$161 - 156$		
		3	**2**	**1**			$321 \cdot 1$	**32**	16
		2	**5**	**9**	3	3	$580 - 321$		
		2	**5**	**8**	**2**	**4**	$3228 \cdot 8$	**322**	161
		(*Rest*)		**1**	**0**	**9**	$933 - 824$	***Root***	**1618**

This shows the **traditional** *square root* **calculation**. We can verify that the **root** *squared* (**1618**^{2} is **2 617 924**) plus the **remainder** (**109**) is the **radicand** (**2 618 033**).

The algorithm is as follows:

We separate from *right to left* and by *two* the digits of the radicand **2 61 80 33**.

We look for *the greatest square less than or equal* to the number of the leftmost radicand not yet used from the previous step (**2**). The *square of* 2 is 4 (exceeds 2) so the **first partial root** is **1**, **whose square is subtracted** from the **2** of the radicand (**2** *minus* **1** is **1**).

We build a number with the previous result (**1**) attached to the left of the next two digits of the radicand (**61**), that is, **1**61.

We double the **partial root** (**1** *times* **2** is **2**); we construct a number with *a digit attached to its right* (**2**6) and *multiply this by that digit* (6); the result (26 *times* 6 is **156**) *must not exceed* **161**.

We find the **partial remainder** by subtracting **161** *minus* **156**. At this point, **261** is **16** *times* **16** plus **5**.

We build a number with the previous result (**5**) attached to the left of the next two digits of the radicand (**80**), that is, **5**80.

We double the **partial root** (**16** *times* **2** is **32**); we construct a number with *a digit attached to its right* (**32**1) and *multiply this by that digit* (1); the product (321 *times* 1 is **321**) *must not exceed* **580** (322 *times* 2 equals 644).

We find the **partial remainder** by subtracting **580** *minus* **321**, namely **259**. At this point, **26 180** is **161** *times* **161** plus **259**.

We build a number with the previous result (**259**) attached to the left of the next two digits of the radicand (**33**), that is, **25 9**33.

We double the **partial root** (**161** *times* **2** is **322**); we construct a number with *a digit attached to its right* (**3 22**8) and *multiply this by that digit* (8); the result of this operation (3 228 *por* 8 son **25 824**) *cannot exceed* **25 933**.

We find the **partial remainder** by subtracting **25 933** *minus* **25 824**, namely **109**. We have reached the end: *the radicand* **2 618 033** *is* **1618** *times* **1618** *plus* **109**.

This algorithm is not the one we will use from now on, but the one devised by Trachtenberg (based on the calculation of the square of a number that we have already seen).

Two-digit numbers

Each of the **square roots** of a **two-digit** number *comes from one of the numbers **0** through **9**.* The **square root** of **0**, **1**, **4**, **9**, **16**, **25**, **36**, **49**, **64**, and **81** is **0**, **1**, **2**, **3**, **4**, **5**, **6**, **7**, **8**, and **9**, respectively.

Three- or four-digit numbers

When calculating the square of a two-digit number, another of at least three figures is generated (because **10** *times* **10** is **100**). Let's take a closer look at **Trachtenberg's method** while we calculate the **square root** of **729**, a three-digit radicand that ***does not produce a remainder*** because *the square of its root*

(27 times 27) is just the radicand:

***Step* 1**. We separate from *right to left* and *two by two* the digits of the radicand (**7 29**).

$\sqrt{7}$	2	9	Calculations	2 7 *Root*
4			$R = 7 - 4$	2^2 $2 \cdot (2 \cdot 7)$ 7^2 (~~04~~ **28** **49**)
3	**0** 2	9	$\mathbf{0} = R - \mathbf{3}$	**3** 2 9
(15)	**2**	**9**	$R = 29 - 29$	~~3~~ **2** **9**
	0	**0**	(R = *Rest*)	*Trachtenberg*

***Step* 2**. We search (by trial and error) for *the greatest square less than or equal to* the number of the radicand on the far left not yet used from the previous step (**7**); 3 *times* 3 is 9 (greater than 7), but 2 *times* 2 is 4 (less than 7). The found digit (**2**) constitutes *the first leftmost digit of the root.*

***Step* 3**. We **subtract** the square of the first digit of the root (2 *times* 2 is **4**) from the one selected from the radicand in the previous step (**7** *minus* **4** is **3**).

***Step* 4**. We take half of the number calculated in the previous step (3 *divided by* 2, truncated is **1**) and *attach* a 0 after it (**10**; however, since 3 *divided by* 2 is more than 1 and less than 2, we better select a number between 10 and 20, namely **15**); we then divide this number by the first digit of the root and truncate this (the truncated division of 15 *by* 2 is **7**). This is **the forecast** for the next digit of the root.

***Step* 5**. With the current answer expected for the root (**27**) we apply the algorithm we saw to square a number:

- We calculate the square of the units (7 *times* 7 is **49**) and write **the two digits** under the units of the root.
- We multiply the units by the tens and double the result (2 *times* 7 is 14; 14 *times* 2 is **28**); **the two digits** are placed under the tens of the root.
- We calculate the square of the tens (2 *times* 2 is **04**); **these two digits** are placed to the left of the previous two numbers under the root, but we

don't need them (~~04~~ **28 49**).

The units of 2**8** collapse with the tens of **4**9 into **2** (**4** *plus* **8** is **12** —**2** units and **1** ***carry***); the carry propagates to the most significant digit of **2**8 (**2** *plus* **1** equals **3**) which thus becomes **3**; as the resulting number (**3**) has only one digit, there is no collapse with the units of 0**4**, which remains invariable and **can be ignored**, *leaving only the three numbers to its right* (**3 2 9**), which we also put under the root. The first of them is subtracted from the current residue (**3**), result (**3** *minus* **3** is **0**) which is attached to the left of the next group of two digits of the radicand not yet used (**2 9**) so that it forms an entity with the most significant digit (**02 9**).

From the number thus formed we subtract the 2 digits to the right of (~~3~~ **2 9**), determining that **00** is the remainder of the square root (**02** *minus* **2** equals **0**; **9** *minus* **9** is **0**).

*In short, the **square root** of 729 is 27 and the **remainder o rest** is 0.*

The **following example** also has a ***three-digit* radicand** (**157**), but this time the **square root** (**12**) is **not exact**, but instead produces a **remainder** (**13**). The example shows that *we must be flexible* when doing the operations to arrive at the correct solution:

$\sqrt{1}$	5	7	Calculations	1	2	*Root*
1			$R = 1 - 1$	1^2 (~~01~~	$2 \cdot (1 \cdot 2)$ **04**	2^2 **04**)
0	**0** 5	7	$\mathbf{0} = R - \mathbf{0}$	**0**	4	4
(00)	**4**	**4**	$R = 57 - 44$	~~0~~	**4**	**4**
	1	**3**	(R = *Rest*)	*Trachtenberg*		

To **solve the square root of 157** we separate its digits from *right to left* and *two by two* (**1 57**); 1 *squared* is 1; the first remainder is 0 (1 *minus* 1); we calculate what the **second digit of the root** can be (0 *divided by* 2 is 0; attach another zero is **0**0; 00 divided by the first digit of the root is 0). In this case it will be necessary to take **at least 1** as *a forecast of the second digit* of the **root** (which would now be **11**); 11 generates the triple

(~~01~~ 02 01) of which we ignore the leftmost number, collapsing the others into 021; the **calculation of the remainder** (05 7 *menos* 2 1) **is 36** (much larger than 11), so *the amount of the second digit must be increased* to 2. **Now the root** is **12**, which produces the triple (~~01~~ 04 04) of which we ignore the leftmost number and the others collapse to 044; the computation of the **remainder** (05 7 *minus* 4 4) **is 13** (somewhat greater than 11); we try to increment by one the second digit of the **root** (now 13); 13 produces the triple (~~01~~ 06 09) of which we ignore the leftmost number and the others collapse to 069; the **remainder** (05 7 *menos* 06 9) would be a negative number, which **is not a viable option**, so *we restore **the root's last valid value*** (**12**). Yet another example is:

$\sqrt{3}$	**4**	**1**	Calculations	**1**	**8**	*Root*
1			$R = 3 - 1$	1^2 (~~01~~	$2 \cdot (1 \cdot 8)$ **16**	8^2 **64**)
2	**0** 4	1	$0 = R - 2$	**2**	2	4
(10)	**2**	**4**	$R = 41 - 24$	~~2~~	**2**	**4**
	1	**7**	(*R* = *Rest*)	*Trachtenberg*		

To **solve for the square root of 341** *we break* 341 into **3 41**; 2 *squared* is 4 (greater than 3); we reduce by 1 (the *square of* 1 is **1** —less than or equal to 3); the **first remainder is 2** (3 *minus* 1); we calculate what **the second digit of the root** can be (2 *divided by* 2 is 1; attaching it a zero is **1**0; 10 divided by the first digit of the root is 10; **we initially take 9**, since it will never be 10). **Now the root** is **19**; this produces the triple (~~01~~ 18 81) which, ignoring the element on the left, collapses to 261 (8 *plus* 8 is 16; the carry of **1**6 plus 1 is 2); the last remainder (2) minus the first digit (2) is 0 (the 2 of ~~2~~61 would be crossed out and 0 would become part of the 4 of 41); **the new remainder is infeasible** (04 1 *minus* 6 1 is negative); we have to *retrace our steps* and *reduce the amount of the second digit* of the **root** (which is **now 18**); this produces the triple (~~01~~ 16 64), which, ignoring the element on the left, collapses to 224; the last remainder (2) minus the first digit (2) is 0 (the 2 is crossed out of ~~2~~24 and the 0 becomes part of the 4 of 41); the remainder or rest is 17 (04 1 *minus* 2 4). To verify the calculation, it is enough to see that ***the square of*** **18** is **324**, plus **17** equals **341**.

Finally, let's see an example with a **four-digit radicand** (**7296**) whose **root is 85** and **the rest is 71**:

$\sqrt{7}$	**2**	**9**	**6**	Calculations	**8 5** *Root*
6	**4**			$\boldsymbol{R} = 72 - 64$	$\mathbf{8^2}$ $2 \cdot (8 \cdot 5)$ 5^2 (~~01~~ **80** **25**)
	8	**0** 9	6	$\mathbf{0} = R - \mathbf{8}$	**8** 2 5
	(40)	**2**	**5**	$\boldsymbol{R} = 96 - 25$	~~8~~ **2** **5**
		7	**1**	(**R** = *Rest*)	*Trachtenberg*

The radicand **7296 is adjusted** *by separating its digits two by two from the left* (**72 96**) to determine that **8 *is the first digit of the root*** (9 *times* 9 exceeds 72). **The first remainder is 8** (72 *minus* 64), which is used to predict the second digit of the root, namely **5** (the quotient 8 *divided by* 2 is 4; after attaching a zero to its right, it is **4**0, and divided by 8 —the first digit of the root— is **5**). The **root is now 85**; using the ***squaring algorithm*** *(tens squared; units times tens, times 2; units squared)* we get the triple (~~01~~ **80 25**) of which the first leftmost element *is ignored* and **the two on the right collapse to 825** *(0 and 2 collapse to their sum, that is,*

2); **the first number** (8) **is subtracted from the current remainder** (8) and the result (0) ***becomes part of the tens*** *of the leftmost digit of the next two not yet used in the radicand (96 becomes **09** 6).* At this point the 8 of ~~8~~25 can be crossed out, leaving 2 5 which we subtract from 09 6 to form the **final remainder** (**71**).

Five- or six-digit numbers

When calculating the square of the maximum number of three figures (999), another one of no more than six is generated (since 999 times 999 is 998 001); but we can determine if the root is going to be two or three digits, just by looking at the number of digits of the radicand and its parity (even or odd):

- If it **is even**, the root will have *exactly* as many digits as *half the digits* of the radicand.
- If it **is odd**, *the number of digits of the radicand is adapted* so that it is even *(adding 1);* the root will consist of *as many digits as half of this result.* E. g. the root of **961** (31 *squared*) has 2 digits (3 *plus* 1 is 4; and *half of* 4 is 2).

To understand the algorithm, there is nothing better than doing it through an example; the radicand (**45 510**) has **five figures**; since 5 is odd, the root will be made up of 3 digits (*add one to* 5 and *divide this by* 2); **45 369** comes from **213** *times* **213**, *plus* **141** equals **45 510**:

$\sqrt{4}$	5	5	1	0	**2 1 3** *Root*
4	*Root* **2**				2^2 $2(2 \cdot 1)$ 1^2 (~~04~~ **04** **01**) **04** **1**
0	**0** 5	**0** 5			$0 - \not{0} = \mathbf{0}$ (*from* **0** 5)
(00)	**4**	**1**			~~**0**~~ **4** **1**
Root **1**	1	4			$1 - \not{1} = \mathbf{0}$ (*from* **0** 5)
	(05)	**2**			$2 \cdot (\mathbf{2} \cdot \mathbf{3}) = \not{1}\ \mathbf{2}$
	Root ~~**2**~~ **3**	2	1	0	210 − 069 = **141**
		0	**6**	**9**	1^2 $2(1 \cdot 3)$ 3^2 (~~01~~ **06** **09**) **06** **9**
Rest =		**1**	**4**	**1**	*Trachtenberg*

We must first *separate the radicand into groups of two digits from right to left* (**4 55 10**). The leftmost group cannot be overtaken by the square of the first digit of the root (in this case 2 *times* 2 equals 4 so **the first digit of the root is 2**).

The **last remainder is 0** (the 4 of the radicand minus the square of 2). To find *the second digit of the root,* we divide 0 by 2, multiply by 10, and divide by the first digit of the root; in this case we select one more for being 0 *—otherwise the next remainder would be 5 and the prediction of the third digit of the root would have more than one digit (5 divided by 2, times 10 equals 25 and half of this is more than 12, greater than 9);* **the second digit of the root is 1**.

Now the root is 21. We square with the method seen above (ignoring the square of the tens) obtaining the triple (~~04~~ 04 01) whose first element is ignored, leaving only the last two: 04 (double the units times tens) and 01 (the square of the units), which collapse to **041** (the tens of 01 with the units of 04). Its first digit (0) is subtracted from the last remainder (0) obtaining a number that is placed to the left of the next digit of the radicand to form an entity (**0**5) from which the second digit of 041 is subtracted (5 *minus* 4 is **1**); **the last remainder is 1**.

To find *the third digit of the root,* we divide 1 by 2, multiply by 10 and divide by the first digit of the root: 1 *divided by* 2, *times* 10 is 05; since 5 *divided by* 2 is between 2 and 3, we choose the largest; **the third digit of the root is 3**.

Now the root is 213. We square with the method seen above that, in addition to the calculations already carried out, requires *multiplying the product of hundreds and units by 2* (2 *times* 3 is 6; *twice* 6 is **12**) and square *13* (the last two digits). The tens (1) of the ~~1~~2 obtained in this paragraph must be subtracted from the last remainder (1) to obtain a number (0) that is placed to the left of the next digit of the radicand and thus form an entity (05) of which we subtract the third digit from 04**1** (5 *minus* 1 is **4**); the units (2) of **12** are subtracted from this value (4 *minus* 2 is **2**); **the last remainder is 2**.

We square 13 (in the same way that we did with 21 at the beginning) obtaining the triple (~~01~~ 06 09) whose first element is ignored, leaving only the last two: 06 y 09, which collapse in **069**; this number must be subtracted from the one formed by the last remainder

(2) attached to the left of the digits of the radicand not yet used (10), that is, **210**, to determine the **remainder or rest** of the square root (210 *menos* 069 is **141**).

Let's explore one more example with a **six-digit** radicand (**245 025**); the root is made up of 3 digits (**495** *squared ends in* **25**; **49** *times* **50** is **2450**).

$\sqrt{2}$	4	5	0	2	5	**4 9 5** *Root*
1	**6**					$\mathbf{4^2}$ $2(4\cdot9)$ 9^2 (~~16~~ **72** **81**) **80** **1**
	8	**0** 5	**1** 0			$8-\mathbf{8}=\mathbf{0}$ (*of* **0** 5)
Root **9**	(40)	**0**	**1**			~~8~~ **0** **1**
		5	9			$5-\mathbf{4}=\mathbf{1}$ (*of* **1** 0)
	Root **5**	(20)	**0**			$2\cdot(\mathbf{4}\cdot\mathbf{5})=\mathbf{4}$ **0**
			9	2	5	$925-925=\mathbf{0}$
			9	**2**	**5**	9^2 $2(9\cdot5)$ 5^2 (~~81~~ **90** **25**) **92** **5**
Remainder =			**0**	**0**	**0**	*Trachtenberg*

The algorithm in detail. The radicand (despite not showing it in the table) is divided from right to left into groups of two elements, that is, **24 50 25**. This is essential for the first step: finding a digit from 1 *to* 9 that squared does not exceed the leftmost group of the radicand (**24**). Since 5 *times* 5 equals 25, this number is **4**, so it is chosen as **the first leftmost digit of the root**. We square 4 (**16**) and subtract this from the leftmost group of the radicand (**24** *minus* **16** is **8** —24 *minus* 14 is 10, *minus* 2 is **8**). Thus, **8** is the ***last remainder*** or ***partial remainder*** (24 is 4 *times* 4 *plus* **8**).

We look for a prediction for the second digit of the root. To do this, we take the *last partial remainder* (**8**), *divide it by* 2 and *attach* a 0 to the right of the result (4 is 8 *divided by* 2 and after attaching it a 0 is **40**); We write it under the remainder in parentheses as a reminder and divide by the first leftmost digit of the root (4); but since 40 *divided by* 4 is 10 (and we are looking for a digit from 0 to 9) **we take 9** as the forecast. Now **the root is 49**.

We use the algorithm seen above **to square 49** (the calculations are on the right of the table under the root) generating a triple whose respective elements are respectively —from left to right— the square of the tens (4 *times* 4 are **16**), twice the product of units and tens (4 *times* 9 is 36; 36 *times* 2 is **72**) and the square of the units (9 *times* 9 is **81**) of which the first of the elements is ignored (actually already has been used) and the remaining two (**72 81**) collapse to **801** —the tens of 81 collapse to the units of 72 (8 *plus* 2 is 10) and the carry of **10** is transferred to the tens of **7**2 (7 *plus* 1 equals 8) producing the pair (80 1)—.

The leftmost **8 is subtracted from the last remainder** (8 *minus* 8 is **0** — it's time to cross out the 8 from ~~8~~01—), a result that is attached to the left most of the next digit of the radicand (that is, to the third from the left), namely, **0**5. **We subtract** the **0** from 8**0**1, from **05** and *use the result* to **predict the third digit of the root**; again, we perform the same calculation as before: 5 *divided by* 2 (truncated) is 2, *times* 10 equals 20 and *divided by* 4 (the first leftmost digit of the root) is 5. **We take 5** as the forecast. Now **the root is 495**.

*Since the root has three digits, finding its square requires additional calculations, namely, twice the product of the extreme digits (4 times 5 times 2 is **40**) and the square of the number formed by the two rightmost* *digits (95 times 95 —the rightmost elements of the triple 81* ***90 25*** *collapse to* ***925****—); these* ***calculations must be added*** *to* ***01*** *(the part of* ~~8~~***01*** *that remains to be used) in such a way that they produce a coherent result that can be subtracted from the radicand and thus generate the* ***remainder or rest*** *of the square root:*

0 1	*Obtained from* ~~**8**~~ **0 1**
4 0	$2\cdot(\mathbf{4}\cdot\mathbf{5})=\mathbf{4\ \ 0}$
+ 9 2 5	$\begin{array}{rccc} & \mathbf{9^2} & 2(9\cdot 5) & 5^2 \\ \mathbf{95^2} = (& \mathbf{\cancel{81}} & \mathbf{90} & \mathbf{25}) \\ & & \mathbf{92} & \mathbf{5} \end{array}$
5 0 2 5	(*Subtracted from the radicand* ***is the rest***)

The operation would end here *(the remainder would be 5025 minus 5025, namely 0);* however, **in our example these calculations are piecemeal applied**.

Let's go back to just when ***5*** *was obtained by subtracting the* ***0*** *from 801, from* ***05*** *and the expected value for* ***the root*** *was* ***495****:*

The first calculation performed is the product of the *extreme digits of* **495** *times* 2, that is, **40**. The 4 of the tens is subtracted from the 5 obtained previously to produce a digit (5 *minus* 4 is **1**) that defines the left end of the next number of the radicand to be considered (0 becomes **1** 0); the 1 of ~~80~~**1** is subtracted from this result (10 *minus* 1 is **9**), which attached to 5 (generated in the difference between **05** and the **0** of 8**0**1) is 59. The 0 of 4**0** is subtracted from the units of 5**9** (9 *minus* 0 is **9**). This 9 is joined to the left of the remaining two digits of the unused radicand (25) to form the numerator (**925**) of the subtraction that will produce the last remainder.

At this point *we calculate* ***the square of*** *the number formed by the last two digits of the root (95)* using the algorithm seen above; but of the resulting triple (~~81~~ 90 25) we ignore the tens' square (81) and the two on the right **collapse to 925** (that is, the tens of **2**5 collapse with the units of 9**0** to **2** —2 *plus* 0 is 2). This is the denominator of the subtraction that will produce the last remainder.

The subtraction of the values calculated in the last two paragraphs constitutes the **remainder** of the square root (**925** *minus* **925** is **0**).

Finally, let's see another example with a *six-digit* radicand; the root will be made up of three digits:

634 *squared* is **401 956**; *plus* **13** is **401 969**. If we group the calculations, the square root looks like this:

$\sqrt{4}$	**0**	**1**	**9**	**6**	**9**	**6 3 4** *Root*
3	**6**	*Root* **6**				$\mathbf{6^2}$ $2(6 \cdot 3)$ 3^2 (~~36~~ **36** **09**) 36 9
Root **3** **(20)**	4					$40 - \mathbf{6^2} = 4$ $(4 \div 2) \cdot 10 = (20)$ $(20 \div 6) = \mathbf{3}$ (*Root*)
Root **4** **(25)**		1	9	6	9	$4 - \mathbf{3} = \mathbf{1}$ (**1** 1) $11 - \mathbf{6} = 5$ $(5 \div 2) \cdot 10 = (25)$ $(25 \div 6) = \mathbf{4}$ (*Root*)
	4	**1**	**9**	**5**	**6**	$2 \cdot (\mathbf{6} \cdot \mathbf{4}) = \mathbf{48}$ 3^2 $2(3 \cdot 4)$ 4^2 (~~09~~ **24** **16**) **25** **6** 3 6 9 + 4 8 + 2 5 6
Rest	**0**	**0**	**0**	**1**	**3**	*Trachtenberg*

The leftmost group of the radicand (**40**) defines **the first digit of the root** (**6**) and the first rest (**4**); with it we determine **the second digit of the root** (**3**), and with the help of the 3 in **3**69, the **third digit of the root** (**4**).

To find the remainder **we add 369** (63^2 ignoring 6^2), **48** (the product of the extremes of the root *by* 2) and **256** (34^2 without the square of 3); the result (**41 956**) must be subtracted from the number formed by the last rest and the digits of the radicand not yet used (**4**1 969), that is, **the remainder is 13**. Partitioning the calculations:

$\sqrt{4}$	0	1	9	6	9	6 3 4 *Root*
3	**6**					6^2 2(6 · 3) 3^2 (~~36~~ **36** **09**) **36** **9**
	4	**1** 1	**1** 9			4 − **3** = **1** (*of* **1** 1)
Root **3**	(20)	**6**	**9**			~~**3**~~ **6** **9**
		5	10			5 − **4** = **1** (*of* **1** 9)
	Root **4**	(25)	**8**			2 · (**6** · **4**) = **4** **8**
			2	6	9	269 − 256 = **13**
			2	**5**	**6**	3^2 2(3 · 4) 4^2 (~~09~~ **24** **16**) **25** **6**
Remainder =			**0**	**1**	**3**	*Trachtenberg*

Seven- or eight-digit numbers

There is a reason why these two cases are treated in the same section: **the square** of the *minimum* **four-digit number** (1000) consists of **seven digits** (1 000 000) and that of the *maximum* **four-digit number** (9999) consists of **eight digits** (99 980 001).

As there are more figures in the root, the number of crossed factors to take into account increases. It's then that hiding the details of the operations that can be performed from memory clarifies the calculation process. E.g., in:

$\sqrt{5}$	4	8	0	2	9	7	**2 3 4 1** *Root*
4	04	18	10	2	9	7	~~1~~ ~~2~~ ~~9~~
							+ ~~1~~ ~~6~~
1	2	1	**10**	**2**	**8**	**1**	+ ~~2~~ **5 6**
							+ **0 4**
(05)	(10)	(05)	**0**	**0**	**1**	**6**	+ **0 6**
							+ **0 8**
Root ~~23~~	*Root* ~~54~~	*Root* ~~21~~		(***R e s t***)			+ **0 1**
		T r a c h t e n b e r g					*Sum:* **1 0 2 8 1**

the calculations of **234 squared** have been simplified:

2^2 $2(2 \cdot 3)$ 3^2 $(\not{0}\not{4}$ 12 $09)$ 12 9	$(23\ squared)$	1 2 9
$2 \cdot (2 \cdot 4) = 16$	$\binom{Cross\ product}{times\ 2}$	1 6
3^2 $2(3 \cdot 4)$ 4^2 $(\not{0}\not{9}$ 24 $16)$ 25 6	$(34\ squared)$	2 5 6

also, the products that **complete the square** of **2341**:

$2 \cdot (2 \cdot 1)$	*The cross products*	0 4
$2 \cdot (3 \cdot 1)$	$(2 \cdot 1), (3 \cdot 1), (4 \cdot 1)$	0 6
$2 \cdot (4 \cdot 1)$	***are doubled,***	0 8
1^2	*the square does not*	0 1

Let's see **the stepwise resolution in detail**:

First, the digits of the radicand must be separated from the right two by two (**5 48 02 97**). The leftmost group (5) allows us to determine **the first digit of the**

root, a number that squared does not exceed it. Since 3 *squared* is 9, the next candidate is **2** and this one does meet the expectations. The root, squared, is 4 (2 *by* 2), which subtracted from the digit of the radicand (5) is **1** (from now on, we will call **last remainder or rest**, the last calculated difference).

To predict **the second digit of the root** we use the last remainder (divided by 2, multiplied by 10 and this divided by the first digit of the root). The *first forecast* is 2 (1 *divided by two, times* 10 equals 05, *divided by* 2 is a number between 2 and 3). **If we take 2 as the second digit of the root**, the square of its now first two digits (**22**) produces the triple (~~04~~ 08 04) of which the last two elements collapse to 084; the 0 of **0**84 must be subtracted from the last remainder (1), a result that is added to the next digit of the radicand to form the entity 04 and from this it would be necessary to subtract 8 (the second digit of 0**8**4), which is not possible without generating a negative number. Thus, *we must reject 2 as a candidate* and choose the next (3). **By taking 3 as the second digit of the root**, the square of the now first 2 digits of the root (**23**) produces the triple (~~04~~ 12 09) with the square of the tens, the double of units and tens, and the units squared, respectively, whose last two elements collapse to **129** (the tens of **0**9 collapse to the units of 1**2**) and the first element is rejected. The **1** of **1**29 must be subtracted from the last remainder (1), a result (**1** – 1 = **0**) that we join to the next digit of the radicand to form the entity **0**4; from this, we subtract the second digit of 1**2**9 to create a **new last remainder** (4 *minus* 2 equals **2**):

$\sqrt{5}$	4	8	0	2	9	7	2 3 4··· *Root*
4	04	18					~~1~~ ~~2~~ 9 ~~1~~ 6 2 5 6
1	2						
(05)	(10)						
Root ~~2~~ 3	*Root* ~~5~~ 4						

To predict **the third digit of the root** we use the last remainder (divided by 2, multiplied by 10 and this divided by the first digit of the root). The **first forecast is 5** (2 *divided by two, times* 10 is 10, *divided by* 2 is **5**). But **if we take 5 as the third digit of the root**, the square of its first three digits (**235**) would be like this:

$\mathbf{2^2}$ $2(2\cdot3)$ 3^2 $(\cancel{04}$ $\mathbf{12}$ $\mathbf{09})$ $\mathbf{12}$ $\mathbf{9}$	$(23\ squared)$	**1 2 9**
$2\cdot(2\cdot5)=\mathbf{20}$	$\begin{pmatrix}Cross\ product\\ \mathbf{times\ 2}\end{pmatrix}$	**2 0**
$\mathbf{3^2}$ $2(3\cdot5)$ 5^2 $(\cancel{09}$ $\mathbf{30}$ $\mathbf{25})$ $\mathbf{32}$ $\mathbf{5}$	$(35\ squared)$	**3 2 5**

To determine the fourth digit of the root, the tens of twice the *cross product* ($\mathbf{2}0$) would be subtracted from the last remainder ($2\ minus\ 2$ is **0**) and the result would be attached to the left of the next digit of the radicand (8) to form an entity ($\mathbf{0}8$) from which to subtract the sum of the third column of the table ($9\ plus\ 0\ plus\ 3$ equals $\mathbf{12}$); but this would produce a negative result ($8\ minus\ 12$), which is not possible. That is why *we must reject 5 as a candidate* and choose a lower value (**4**). **By taking 4 as the third digit of the root**, the square of its first three digits (**234**) already collapsed is (ignoring the square of 2):

1 2 9

1 6

2 5 6

To determine **the fourth digit of the root** we subtract the tens of twice the *cross product* (**16**) from the last remainder (2 *minus* 1 is **1**), attaching the result to the left of the next digit of the radicand (8) to form an entity (**18**) from which to subtract the sum of the third column of the previous table (9 *plus* 6 *plus* 2 equals **17**); the result of this operation (18 *minus* 17 is **1**) **determines the new last remainder**.

The *prediction of the fourth digit of the root* requires the last remainder (divided by 2, multiplied by 10 and divided by the first digit of the root). The *first forecast* is 2 (1 *divided by* 2, *times* 10 is 05; *divided by* 2 is a number between 2 and 3). **If we take 2 as the fourth digit of the root**, the square of its now four digits (**2342**) would be completed with the products:

$\mathbf{2 \cdot (2 \cdot 2)}$	*The cross products*	**0**	**8**			
$\mathbf{2 \cdot (3 \cdot 2)}$	$(\mathbf{2 \cdot 2}), (\mathbf{3 \cdot 2}), (\mathbf{4 \cdot 2})$	+	**1**	**2**		
$\mathbf{2 \cdot (4 \cdot 2)}$	***are doubled,***		+	**1**	**6**	
$\mathbf{2^2}$	*the square does not*			+	**0**	**4**

Of all of them, it would be necessary to subtract the tens from the first above (**08**) from the last remainder (1) and attach the result (1) to the next digit of the radicand (0) to form a number (**10**), which together with the digits not yet used from the radicand form **a new last remainder** (**10 297**) from which to subtract the following sum (to obtain the remainder of the square root):

~~1~~	~~2~~	~~9~~				
	~~1~~	~~6~~				
		~~2~~	5	6		
	+	~~0~~	8			
		+	1	2		
			+	1	6	
				+	0	4
–	–	–	–	–	–	–
		1	**4**	**9**	**6**	**4**

But 10 297 *minus* 14 964 is negative, so **the fourth digit of the root** must be one minus (**1**):

√5	4	8	0	2	9	7
4	04	18	10			
1	2	1				
(05)	(10)	(05)				
Root **~~2~~ 3**	*Root* **~~5~~ 4**	*Root* **~~2~~ 1**				

2 3 4 1 *Root*

~~1~~	**~~2~~**	**~~9~~**			
+ **~~1~~**	**~~6~~**				
	+ **~~2~~**	**5**	**6**		
	+ **0**	**4**			
		+ **0**	**6**		
			+ **0**	**8**	
				+ **0**	**1**

Taking 1 as the fourth digit of the root, the square of its four digits (**2341**) must be completed like this:

$\mathbf{2 \cdot (2 \cdot 1)}$	*The cross products*	**0 4**
$\mathbf{2 \cdot (3 \cdot 1)}$	$(\mathbf{2} \cdot \mathbf{1}), (\mathbf{3} \cdot \mathbf{1}), (\mathbf{4} \cdot \mathbf{1})$	**0 6**
$\mathbf{2 \cdot (4 \cdot 1)}$	***are doubled,***	**0 8**
$\mathbf{1^2}$	*the square does not*	**0 1**

As before, we need to subtract the tens of **0**4 from the last remainder and join the result (1 *minus* 0 is **1**) to the next digit of the radicand (0) to form the entity (**10**), which we join to the remaining digits of the radicand to form **a new last residue** (**10 297**) from which to subtract the following sum (and thus find the remainder of the square root):

```
1̶  2̶  9̶
   1̶  6̶
      2̶  5  6
   +  0̶  4
      +  0  6
         +  0  8
            +  0  1
–  –  –  –  –  –  –
      1  0  2  8  1
```

The **remainder** is **16** (10 297 *minus* 10 281) and $\mathbf{2341^2}$ (the root, squared) *plus* **16** is **5 480 297** (the radicand).

The **square root** of an **eight-digit radicand** is similar. The four digits of the root are predicted as before until we obtain them all, at which point we can make the calculations of the products whose sum (from right to left and considering the carries) allows us to find the rest of the square root:

√4	**2**	**6**	**6**	**7**	**1**	**5**	**7**	**6 5 3 2** *Root*
3	**6**	06	16	07	1	5	7	~~6~~ ~~2~~ ~~5~~
								+ ~~3~~ ~~6~~
	6	4	2	**7**	**0**	**2**	**4**	+ ~~3~~ **0 9**
								+ ~~2~~ **4**
	(30)	(20)	(10)	**0**	**1**	**3**	**3**	+ **2 0**
								+ **1 2**
	Root **5**	*Root* **3**	*Root* ~~1~~ **2**		(***R e s t***)			+ **0 4**
	Trachtenberg							*Sum*: **7 0 2 4**

The **calculation is correct**; the radicand **42 667 157** is 42 667 024 (the root **6532** *squared*) *plus* **133** (the rest).

Likewise, we can subtract "the sum of each column of the products that make up the square of 6532" from "each digit of the radicand" *(from left to right),* taking the result of each difference to the immediate column to its right:

$\sqrt{42}$	6	6	7	1	5	7	6 5 3 2 *Root*
36	**0**	**1**	**0**	**1**	**1**	**13**	~~6~~ ~~2~~ ~~5~~
	6	6	7	1	5	7	+ ~~3~~ ~~6~~
6	4	2	**6** (−)	**10** (−)	**2** (−)	**4** (−)	+ ~~3~~ **0 9** + ~~2~~ **4**
Root **5**	*Rt* **3**	*Rt* **2**			***Rest*** =	**133**	+ **2 0** + **1 2** + **0 4**
Trachtenberg							**6 10 2 4** *Columns' sum*

This would have to be done once the fourth digit of the root has been determined, at which point the products that complement the square of said root are calculated; then we must add by columns the elements not yet used: **6** (0 *plus* 4 *plus* 2), **10** (9 *plus* 0 *plus* 1), **2** (2 *plus* 0) and **4**, from left to right, as they are subtracted from each digit of the radicand still pending use (from left to right) so that each subtraction produces a result that is transferred attached to the left of the numerator of the following operation, that is, the **6** is subtracted from the **0**7 of the radicand giving **1**; this is attached to the left of the next digit of the radicand (**1**1) and the sum of the next column (**10**) is subtracted from it, giving **1**; result that attached to the left of the **5** of the radicand (**1**5), minus the sum of the next column (**2**), is **13**; which is transferred to the tens of the **7** of the radicand (**13**7); value that, after subtracting from it the last sum of columns (**4**) defines the rest of the square root (**133**).

Numbers with nine or more digits

The principle followed when calculating the square root of a number of nine or more digits is similar to cases of seven or eight digits. We only have to pay attention to the fact that *for each number added to the root we must add its square and the double of each one of the duly displaced crossed products.*

To see this, let's first square **2**, **23**, **234**, **2345** and **23456** building on each previous calculation.

The *square of* **2** is **4** (a single addend). The rest of the squares first provide the addends corresponding to **the double of each *cross product*** (2 *times* 3, 2 *times* 4, 2 *times* 5 and/or 2 *times* 6) and then **the square of the**

new digit (2, 3, 4, 5 ó 6), as the case may be, displaced as follows:

$$
\begin{array}{r@{\;}l ccccccccc}
\mathbf{2}^2 & = & \mathbf{4} & & & & & & & & \\
2\cdot(\mathbf{2}\cdot 3) & = & \mathbf{1} & \mathbf{2} & & & & & & & \\
\mathbf{3}^2 & = & & \mathbf{0} & \mathbf{9} & & & & & & \\
 & + & - & - & - & & & & & & \\
\mathbf{23}^2 & = & 5 & 2 & 9 & & & & & & \\
2\cdot(\mathbf{2}\cdot 4) & = & & 1 & 6 & & & & & & \\
2\cdot(\mathbf{3}\cdot 4) & = & & & 2 & 4 & & & & & \\
\mathbf{4}^2 & = & & & & 1 & 6 & & & & \\
 & + & - & - & - & - & - & & & & \\
\mathbf{234}^2 & = & \mathbf{5} & \mathbf{4} & \mathbf{7} & \mathbf{5} & \mathbf{6} & & & & \\
2\cdot(\mathbf{2}\cdot 5) & = & & & \mathbf{2} & \mathbf{0} & & & & & \\
2\cdot(\mathbf{3}\cdot 5) & = & & & & \mathbf{3} & \mathbf{0} & & & & \\
2\cdot(\mathbf{4}\cdot 5) & = & & & & & \mathbf{4} & \mathbf{0} & & & \\
\mathbf{5}^2 & = & & & & & & \mathbf{2} & \mathbf{5} & & \\
 & + & - & - & - & - & - & - & - & & \\
\mathbf{2345}^2 & = & 5 & 4 & 9 & 9 & 0 & 2 & 5 & & \\
2\cdot(\mathbf{2}\cdot 6) & = & & & & 2 & 4 & & & & \\
2\cdot(\mathbf{3}\cdot 6) & = & & & & & 3 & 6 & & & \\
2\cdot(\mathbf{4}\cdot 6) & = & & & & & & 4 & 8 & & \\
2\cdot(\mathbf{5}\cdot 6) & = & & & & & & & 6 & 0 & \\
\mathbf{6}^2 & = & & & & & & & & 3 & 6 \\
 & + & - & - & - & - & - & - & - & - & - \\
\mathbf{23456}^2 & = & \mathbf{5} & \mathbf{5} & \mathbf{0} & \mathbf{1} & \mathbf{8} & \mathbf{3} & \mathbf{9} & \mathbf{3} & \mathbf{6}
\end{array}
$$

In each section (between the squares of 2, 23, 234, 2345 and 23456), we can take advantage of the calculations already made and start the sum from the last square. For example, we can calculate the square of **234 567** from

550 183 936 (the square of 23 456) by simply adding each of the cross products (2 *times* 7, 3 *times* 7, 4 *times* 7, 5 *times* 7 and 6 *times* 7) multiplied by 2, and the *square of* 7, *properly displaced:*

$\mathbf{23456}^2 =$	5	5	0	1	8	3	9	3	6		
$2 \cdot (\mathbf{2} \cdot 7) =$					2	8					
$2 \cdot (\mathbf{3} \cdot 7) =$						4	2				
$2 \cdot (\mathbf{4} \cdot 7) =$							5	6			
$2 \cdot (\mathbf{5} \cdot 7) =$								7	0		
$2 \cdot (\mathbf{6} \cdot 7) =$									8	4	
$\mathbf{7}^2 =$										4	9
+	–	–	–	–	–	–	–	–	–	–	–
$\mathbf{234567}^2 =$	**5**	**5**	**0**	**2**	**1**	**6**	**7**	**7**	**4**	**8**	**9**

To **place these addends** properly:

- The units of the square of the new root digit must match those of the result.
- The doubles of the cross products are placed in order (84, 70, 56, 42 y 28) from bottom to top and shifted to the left so that their units match the tens of the previous.

Dividing the radicand into groups of 2 determines the digits of the root. E.g., the square root of **5 50 18 39 36** is

23456 (five digits, as many as groups):

$\sqrt{5}$	5	0	1	8	3	9	3	6	**2 3 4 5 6**
4	**0**	**2**	**1**						~~1~~ ~~2~~ ~~9~~
	5	0	1						+ ~~1~~ ~~6~~
1	3	3	3	8	3	9	3	6	+ ~~2~~ ~~5~~ 6
									+ ~~2~~ ~~0~~
			3	**8**	**3**	**9**	**3**	**6**	+ ~~3~~ 0
									+ 4 0
Remainder =			**0**	**0**	**0**	**0**	**0**	**0**	+ 2 5

Trachtenberg		*Sum* = **1025**
Operations	***Actions***	**2** · (*Cross prod. of* **6**) *and* $\mathbf{6}^2$
$5 - 4 = 1$	*Subtract* 2^2	
$1 - 1 = 0$	~~1~~29	**1 0 2 5**
0$5 - 2 = 3$	~~12~~9	**2 4**
$3 - 1 = 2$	~~1~~6	+ **3 6**
$9 + 6 + 2 = 17$	*Add column*	+ **4 8**
2$0 - 17 = 3$	~~129~~ ~~16~~ ~~2~~56	+ **6 0**
$3 - 2 = 1$	~~2~~0	+ **3 6**
$5 + 0 + 3 = 8$	*Add column*	– – – – – –
$11 - 8 = 3$	~~25~~6 ~~20~~ ~~3~~0	**3 8 3 9 3 6**

Chapter 7

Cubes and its roots

If we represent a number N by a straight line of equal length to its magnitude, *we can add N of those lines (each one adjacent to the previous one) to form a surface*. Its area is N *times* N and it turns out to be a square. *The product* **N** *times* **N** *is called* ***the square of*** *a number*. If we put ***N*** **squares** on top of each other, the area of the figure thus created is ***N*** *times* ***N*** *times* ***N*** *and the product is called* ***the cube of*** *a number*.

The *inverse operation* of the cube of a number is ***the cube root***, which consists of *finding the number N that when cubed gives as a result N times N times N*. **E.g.**, **the cube of 7** is **343**, so **7** is its **cube root**.

Cubic root of a number

If we take the number $\boldsymbol{N}$ and multiply it by itself three times we get $\boldsymbol{N \cdot N \cdot N}$ *(that is, we have cubed it).* The **cube root** of **the radicand** $\boldsymbol{N \cdot N \cdot N}$ is $\boldsymbol{N}$; all this is denoted like this: $\sqrt[3]{\boldsymbol{N \cdot N \cdot N}} = \sqrt[3]{\boldsymbol{N}^3} = \boldsymbol{N}^{\frac{3}{3}} = \boldsymbol{N}$.

$\sqrt[3]{\mathbf{4}}$	**2**	**1**	**8**	**9**	**6**	**7** $\underline{\mathbf{5}}$ *Root*
3	**4**	**3**	*Root* **7**			$7^3 = \mathbf{343}$
Root **5**	$\boxed{7}$	$\boxed{8}$	$\boxed{8}$	9	6	$3 \cdot 7^2 = \mathbf{147}$ $\boxed{788} \div \mathbf{147} = \underline{\mathbf{5}}$
	7	**3**	**5**			$3 \cdot (7^2 \cdot 5) = \mathbf{735}$
		5	**2**	**5**		$3 \cdot (7 \cdot 5^2) = \mathbf{525}$
			1	**2**	**5**	$5^3 = \mathbf{125}$
	0	0	0	**2**	**1**	*(remainder or rest)*

This is what the **traditional** *cube root* **calculation** looks like. The **root**, cubed (75 *times* 75 *times* 75 is **421 875**), plus the **remainder** (**21**), is the **radicand** (**421 896**).

The algorithm is as follows:

We separate *from right to left and three by three* the digits of the radicand **421 896**.

We look for *the largest cube less than or equal* to the number of the leftmost radicand not yet used from the previous step (**421**). the *cube of* 8 is 512 (which exceeds 421) so **the first partial root** is **7**; **its cube must be subtracted** from the **421** of the radicand (**421** *minus* **343** equals **78**).

We attach to the right of the previous rest (78), the following three numbers of the radicand (896) and from the resulting number we separate two digits from the right (**788 96**); with the remaining leftmost digits we estimate the next digit of the root (dividing it by *triple the square of the root*—**147**). The prediction of the second digit of the root is **5** (788 *divided by* 147).

To **obtain the remainder**, from the *last remainder* (78 896) *we must subtract* ***three numbers*** *duly shifted from left to right:*

For **the first**, we have **to square the number formed by the first digits of the root up to the current one, by the current one and triple the result** ($7\ squared, times$ 5 is 245; $times$ 3 is **735**); value that we must align from the left with the most significant digit of the last remainder.

For **the second**, we have to calculate **the square of the current digit of the root by the number formed by the first digits of the root up to the current one** (those on its left) **and triple the result** ($5^2 \cdot 7$ equals 175; $times$ 3 is **525**); value that is shifted one position to the right from the previous one.

For **the third**, we find **the cube of the current root digit** (5 $cubed$ is 125) and shift it one position to the right from the previous one.

The three *duly displaced* **add up to** 78 875, which subtracted from the last remainder (78 896) constitutes **the rest** of the cube root (78896 $minus$ 78875 is **21**).

To check the calculations: **75** (the **root**) *cubed* is **421 875** $plus$ **21** (the **rest**) is **421 896** (the **radicand**).

Let's look at another example. This time the radicand is the exact cube of **123**, namely **1 860 867**.

$\sqrt[3]{1}$	8	6	0	8	6	7	1 2̲ 3̲ *Root*
1	*Root* **1**						$1^3 = 1$
[0]	[8]	6	0	*Root* **2**			$3 \cdot 1^2 = 3$ $\boxed{08} \div 3 = \underline{2}$
0	**6**						$3 \cdot (1^2 \cdot 2) = 06$
	1	**2**					$3 \cdot (1 \cdot 2^2) = 12$
		0	**8**				$2^3 = 08$
Root **3**	[1]	[3]	[2]	[8]	6	7	$3 \cdot 12^2 = 432$ $\boxed{1328} \div 432 = \underline{3}$
	1	**2**	**9**	**6**			$3 \cdot (12^2 \cdot 3)$
			3	**2**	**4**		$3 \cdot (12 \cdot 3^2)$
					2	**7**	$3^3 = 27$
	0	0	0	0	0	0	*(R e s t)*

The observant reader will have already seen that for

each new digit in the root we only need to consider the three numbers mentioned above:

- The square of the number formed by the digits of the root to its left, by the digit itself, *times* 3.
- The number formed by the digits of the root to its left, by the square of the digit itself, *times* 3.
- The cube of the digit itself.

To fix ideas, nothing better than looking at the cubes of **2**, **23**, **234** and **2345**, taking advantage of each previous calculation:

$$
\begin{array}{rcccccccccc}
\mathbf{2}^3 = & \mathbf{8} & & & & & & & & & \\
3\cdot(\mathbf{2}^2\cdot\mathbf{3}) = & \mathbf{3} & \mathbf{6} & & & & & & & & \\
3\cdot(\mathbf{2}\cdot\mathbf{3}^2) = & & \mathbf{5} & \mathbf{4} & & & & & & & \\
\mathbf{3}^3 = & & & \mathbf{2} & \mathbf{7} & & & & & & \\
+ & - & - & - & - & & & & & & \\
\mathbf{23}^3 = & 12 & 1 & 6 & 7 & & & & & & \\
3\cdot(\mathbf{23}^2\cdot\mathbf{4}) = & & 6 & 3 & 4 & 8 & & & & & \\
3\cdot(\mathbf{23}\cdot\mathbf{4}^2) = & & & 1 & 1 & 0 & 4 & & & & \\
\mathbf{4}^3 = & & & & & & 6 & 4 & & & \\
+ & - & - & - & - & - & - & - & & & \\
\mathbf{234}^3 = & \mathbf{12} & \mathbf{8} & \mathbf{1} & \mathbf{2} & \mathbf{9} & \mathbf{0} & \mathbf{4} & & & \\
3\cdot(\mathbf{234}^2\cdot\mathbf{5}) = & & & \mathbf{8} & \mathbf{2} & \mathbf{1} & \mathbf{3} & \mathbf{4} & \mathbf{0} & & \\
3\cdot(\mathbf{234}\cdot\mathbf{5}^2) = & & & & & \mathbf{1} & \mathbf{7} & \mathbf{5} & \mathbf{5} & \mathbf{0} & \\
\mathbf{5}^3 = & & & & & & & & \mathbf{1} & \mathbf{2} & \mathbf{5} \\
+ & - & - & - & - & - & - & - & - & - & - \\
\mathbf{2345}^3 = & 12 & 8 & 9 & 5 & 2 & 1 & 3 & 6 & 2 & 5
\end{array}
$$

We can obtain another way to solve the cube root more in accordance with the Trachtenberg algorithms. The number **321** can help us to investigate *the relationship between its cube and each of the digits that make up its square*, simply by multiplying in the traditional way:

$$
\begin{array}{ccccccc}
 & & & & \mathbf{3} & \mathbf{2} & \mathbf{1} \\
 & & & \times & \mathbf{3} & \mathbf{2} & \mathbf{1} \\
 & & & - & - & - & - \\
 & & & & 1\cdot 3 & 1\cdot 2 & 1\cdot 1 \\
 & & & 2\cdot 3 & 2\cdot 2 & 2\cdot 1 & + \\
 & & 3\cdot 3 & 3\cdot 2 & 3\cdot 1 & + & \\
 & & - & - & - & - & - \\
 & & \boldsymbol{e} & \boldsymbol{d} & \boldsymbol{c} & \boldsymbol{b} & \boldsymbol{a} \\
 & & & \times & \mathbf{3} & \mathbf{2} & \mathbf{1} \\
 & & - & - & - & - & - \\
 & & 1e & 1d & 1c & 1b & 1a \\
 & 2e & 2d & 2c & 2b & 2a & + \\
3e & 3d & 3c & 3b & 3a & + & \\
- & - & - & - & - & - & - \\
\mathbf{3}\boldsymbol{e} & \mathbf{2}\boldsymbol{e} & \mathbf{1}\boldsymbol{e} & \mathbf{1}\boldsymbol{d} & \mathbf{1}\boldsymbol{c} & \mathbf{1}\boldsymbol{b} & \mathbf{1}\boldsymbol{a} \\
 & + & + & + & + & + & \\
 & \mathbf{3}\boldsymbol{d} & \mathbf{2}\boldsymbol{d} & \mathbf{2}\boldsymbol{c} & \mathbf{2}\boldsymbol{b} & \mathbf{2}\boldsymbol{a} & \\
 & & + & + & + & & \\
 & & \mathbf{3}\boldsymbol{c} & \mathbf{3}\boldsymbol{b} & \mathbf{3}\boldsymbol{a} & &
\end{array}
$$

Then the *cube of* **32** can be obtained from its square (calculated using Trachtenberg) like this:

$$
\begin{array}{rccccc}
\mathbf{3}^2 = & \mathbf{9} & & & \\
2\cdot(\mathbf{3}\cdot 2) = & \mathbf{1} & \mathbf{2} & & \\
\mathbf{2}^2 = & & & \mathbf{4} & \\
+ & c & b & a & \\
\mathbf{32}^2 = & 10 & 2 & 4 & \\
(\mathbf{3}-1)c = & 20 & & & \\
(\mathbf{3}-1)b+\mathbf{2}c = & 2 & 4 & & \\
(\mathbf{3}-1)a+\mathbf{2}b = & & 1 & 2 & \\
\mathbf{2}a = & & & 0 & 8 \\
+ & - & - & - & - \\
\mathbf{32}^3 = & \mathbf{32} & \mathbf{7} & \mathbf{6} & \mathbf{8}
\end{array}
$$

And the *cube of* **321** as follows:

$$
\begin{array}{rccccccc}
\mathbf{3}^2 = & \mathbf{9} & & & & & & \\
2\cdot(\mathbf{3}\cdot 2) = & \mathbf{1} & \mathbf{2} & & & & & \\
\mathbf{2}^2 = & & & \mathbf{4} & & & & \\
+ & - & - & - & & & & \\
\mathbf{32}^2 = & 10 & 2 & 4 & & & & \\
2\cdot(\mathbf{3}\cdot 1) = & & 0 & 6 & & & & \\
2\cdot(\mathbf{2}\cdot 1) = & & & 0 & 4 & & & \\
\mathbf{1}^2 = & & & & 0 & 1 & & \\
+ & e & d & c & b & a & & \\
\mathbf{321}^2 = & \mathbf{10} & \mathbf{3} & \mathbf{0} & \mathbf{4} & \mathbf{1} & & \\
(\mathbf{3}-1)e = & \mathbf{20} & & & & & & \\
(\mathbf{3}-1)d+\mathbf{2}e = & \mathbf{2} & \mathbf{6} & & & & & \\
(\mathbf{3}-1)c+\mathbf{2}d+\mathbf{1}e = & & \mathbf{1} & \mathbf{6} & & & & \\
(\mathbf{3}-1)b+\mathbf{2}c+\mathbf{1}d = & & & \mathbf{1} & \mathbf{1} & & & \\
(\mathbf{3}-1)a+\mathbf{2}b+\mathbf{1}c = & & & & \mathbf{1} & \mathbf{0} & & \\
\mathbf{2}a+\mathbf{1}b = & & & & & \mathbf{0} & \mathbf{6} & \\
\mathbf{1}a = & & & & & & \mathbf{0} & \mathbf{1} \\
+ & - & - & - & - & - & - & - \\
\mathbf{321}^3 = & 33 & 0 & 7 & 6 & 1 & 6 & 1
\end{array}
$$

where the digit **1** must be understood as the units of 32**1** except in (**3** – 1) which means *subtract* 1 from the hundreds of **3**21; the digit **2** represents the tens of 3**2**1; and the digit **3**, the hundreds of **3**21; In this way, it is possible to generalize and obtain an algorithm to solve the cube root of a number.

But let's see first how to convert *the square* of 321 into *the square* of 321**5**, and this into *a cube:*

$$
\begin{array}{rcllllllllll}
\mathbf{321}^2 & = & 10 & 3 & 0 & 4 & 1 & & & & & \\
2\cdot(\mathbf{3}\cdot 5) & = & & & 3 & 0 & & & & & & \\
2\cdot(\mathbf{2}\cdot 5) & = & & & & 2 & 0 & & & & & \\
2\cdot(\mathbf{1}\cdot 5) & = & & & & & 1 & 0 & & & & \\
\mathbf{5}^2 & = & & & & & & 2 & 5 & & & \\
 & + & - & - & - & - & - & - & - & & & \\
\mathbf{3215}^2 & = & \mathbf{10} & \mathbf{3} & \mathbf{3} & \mathbf{6} & \mathbf{2} & \mathbf{2} & \mathbf{5} & & & \\
 & & g & f & e & d & c & b & a & & & \\
(\mathbf{3}-1)g & = & \mathbf{20} & & & & & & & & & \\
(\mathbf{3}-1)f+\mathbf{2}g & = & \mathbf{2} & \mathbf{6} & & & & & & & & \\
(\mathbf{3}-1)e+\mathbf{2}f+\mathbf{1}g & = & & \mathbf{2} & \mathbf{2} & & & & & & & \\
(\mathbf{3}-1)d+\mathbf{2}e+\mathbf{1}f+\mathbf{5}g & = & & & \mathbf{7} & \mathbf{1} & & & & & & \\
(\mathbf{3}-1)c+\mathbf{2}d+\mathbf{1}e+\mathbf{5}f & = & & & & \mathbf{3} & \mathbf{4} & & & & & \\
(\mathbf{3}-1)b+\mathbf{2}c+\mathbf{1}d+\mathbf{5}e & = & & & & & \mathbf{2} & \mathbf{9} & & & & \\
(\mathbf{3}-1)a+\mathbf{2}b+\mathbf{1}c+\mathbf{5}d & = & & & & & & \mathbf{4} & \mathbf{6} & & & \\
\mathbf{2}a+\mathbf{1}b+\mathbf{5}c & = & & & & & & & \mathbf{2} & \mathbf{2} & & \\
\mathbf{1}a+\mathbf{5}b & = & & & & & & & & \mathbf{1} & \mathbf{5} & \\
\mathbf{5}a & = & & & & & & & & & \mathbf{2} & \mathbf{5} \\
\mathbf{3215}^3 & = & 33 & 2 & 3 & 0 & 9 & 6 & 3 & 3 & 7 & 5
\end{array}
$$

Finally, the **example**:

$\sqrt[3]{4}$	**2**	**1**	**8**	**9**	**6**		$\underline{7}\ \underline{5}$ *Root*
						Root **7**	$7^3 = 343$ $421 - 343 = \boxed{78}$
	5	6	2	5		*Root* **5**	$\underline{7}^2 = 49$ $2(\underline{7} \cdot \underline{5}) = 7\ 0$ $\underline{5}^2 = \ \ 2\ 5$ $+\ c\ \ b\ \ a$ $\underline{75}^2 = \mathbf{56}\ \mathbf{2}\ \mathbf{5}$
3	3	6					$(\underline{7} - 1)c = 6 \cdot \mathbf{56}$
	2	9	2				$(\underline{7} - 1)b + \underline{5}c$
		0	4	0			$(\underline{7} - 1)a + \underline{5}b$
			0	2	5		$\underline{5}a = \underline{5} \cdot 5$
0	0	0	0	**2**	**1**		*(Remainder)*

Initially, the radicand is divided into groups of three digits from right to left (**421 896**). This makes it easy to determine what number cubed does not exceed the leftmost group of the radicand (421). Since 8^3 is 512, we choose **7 as the first digit of the root**. We *square* 7 and **look for the next two-digit number** such that **when converting its square into a cube** it does not exceed the leftmost digits of the radicand. For this we use the Trachtenberg algorithm (although any other method can be chosen); to 7^2 we add (duly displaced) the *double of the product* «7 *times* ***the test number***» ($7 \cdot \mathbf{5} = 35$; $35 \cdot 2 = 70$) and *the square* of *it* ($\mathbf{5}^2 = 25$), determining that 75^2 is 5625; we denote the units $\boldsymbol{a}$, the tens $\boldsymbol{b}$ and the rest $\boldsymbol{c}$. To **convert the square into a cube** (75^3) it is enough to add properly displaced, in order from top to bottom and from left to right $(D-1)\boldsymbol{c}$, $(D-1)\boldsymbol{b}+U\boldsymbol{c}$, $(D-1)\boldsymbol{a}+U\boldsymbol{b}$ and $U\boldsymbol{a}$, where in this case U are the units of $7\mathbf{5}$ and D its tens. The result is 421 875, which subtracted from 421 896 (the radicand) gives us the remainder of the cube root (21).

If there are still digits of the radicand unused, we must start from the last square found and find a square with one more figure that, when converted into a cube, does not exceed the value of the radicand.

Chapter 8

Fourth root or higher

Denoting N the number that we are going to consider, raising N to the *fourth power* is multiplying it *four times by itself* $(N \cdot N \cdot N \cdot N)$ and is denoted as N^4.

The fourth root of N^4 is N, since it is the number that must be raised to the fourth power to obtain N^4:

$$\sqrt[4]{N^4} = (N^4)^{\frac{1}{4}} = N^{4 \cdot \frac{1}{4}} = N^{\frac{4}{4}} = N^1 = N$$

Also, *the fourth root of a number* consists of calculating *the square root of its square root*, that is:

$$\sqrt[4]{N} = N^{\frac{1}{4}} = N^{\frac{1}{2} \cdot \frac{1}{2}} = \left(N^{\frac{1}{2}}\right)^{\frac{1}{2}} = \left(\sqrt[2]{N}\right)^{\frac{1}{2}} = \sqrt[2]{\sqrt[2]{N}}$$

E.g., the $fourth\ root\ of$ 625 is 5 (since $\mathbf{5}$ $times$ $\mathbf{5}$ is 25, $times$ $\mathbf{5}$ is 125, and by $\mathbf{5}$ is 625); the $square\ root\ of$ 625 is

25 (25 *times* 25 is 625) and the *square root of* 25 is 5 (because 5 *times* 5 equals 25).

We can obtain another way to solve the cube root more in accordance with the Trachtenberg algorithms. The terms to add to convert the *cube of* **31** to its fourth power are shown here:

$$
\begin{array}{rcccc}
\mathbf{3}^2 = & 9 & & & \\
2\cdot(\mathbf{3}\cdot 1) = & 0 & 6 & & \\
\mathbf{1}^2 = & & & 1 & \\
+ & - & - & - & \\
\mathbf{31}^2 = & \mathbf{9} & \mathbf{6} & \mathbf{1} & \\
(\mathbf{3}-1)\cdot 9 = & \mathbf{18} & & & \\
(\mathbf{3}-1)\cdot 6+\mathbf{1}\cdot 9 = & \mathbf{2} & \mathbf{1} & & \\
(\mathbf{3}-1)\cdot 1+\mathbf{1}\cdot 6 = & & \mathbf{0} & \mathbf{8} & \\
\mathbf{1}\cdot 1 = & & & \mathbf{0} & \mathbf{1} \\
+ & - & - & - & - \\
\mathbf{31}^3 = & 29 & 7 & 9 & 1 \\
\boxed{cba}\, pattern\ (over\ 31^3) & \boxed{\cancel{29}} & \boxed{\cancel{7}} & \boxed{\cancel{9}} & \\
Pattern\ reset\ \boxed{cba} = & \boxed{297} & \boxed{9} & \boxed{1} & \\
(\mathbf{3}-1)c = & 594 & & & \\
(\mathbf{3}-1)b+\mathbf{1}c = & 31 & 5 & & \\
(\mathbf{3}-1)a+\mathbf{1}b = & & 1 & 1 & \\
\mathbf{1}a = & & & 0 & 1 \\
+ & - & - & - & - \\
\mathbf{31}^4 = & \mathbf{923} & \mathbf{5} & \mathbf{2} & \mathbf{1}
\end{array}
$$

Adjusting to the number of digits available (the units

and tens of 31) can save additional steps. With the readjustment (and from bottom to top) the first addend uses the letter $\boldsymbol{a}$ next to the units of 3**1**, the second the letters $\boldsymbol{a}$ and $\boldsymbol{b}$ next to the tens and units of **31**, the third the letters $\boldsymbol{b}$ and $\boldsymbol{c}$ next to the tens and units of **31**, and the fourth the letter $\boldsymbol{c}$ next to the tens of **3**1. If we wish, *each letter in the pattern can be set to act only on single digits* of 31^3, in which case the pattern of letters ba must be shifted from left to right (over 29 791) in several steps as follows:

$$
\begin{array}{r cccccc}
 & c & b & a & & & \\
\mathbf{31}^3 = & \boxed{2} & \boxed{9} & \boxed{7} & 9 & 1 & \\
(\mathbf{3}-1)c = & 4 & & & & & \\
(\mathbf{3}-1)b+\mathbf{1}c = & 2 & 0 & & & & \\
(\mathbf{3}-1)a+\mathbf{1}b = & & 2 & 3 & & & \\
+ & - & - & - & - & - & \\
 & \mathbf{9} & \mathbf{2} & \mathbf{0} & \mathbf{9} & \mathbf{1} & \\
\boxed{ba}\,pattern\ (over\ 31^3) & * & * & \boxed{7} & \boxed{9} & * & \\
(\mathbf{3}-1)a+\mathbf{1}b = & & & \mathbf{2} & \mathbf{5} & & \\
+ & - & - & - & - & - & \\
 & 9 & 2 & 3 & 4 & 1 & \\
\boxed{ba}\,pattern\ (over\ 31^3) & * & * & * & \boxed{9} & \boxed{1} & \\
(\mathbf{3}-1)a+\mathbf{1}b = & & & & 1 & 1 & \\
\mathbf{1}a = & & & & & 0 & 1 \\
+ & - & - & - & - & - & - \\
\mathbf{31}^4 = & \mathbf{9} & \mathbf{2} & \mathbf{3} & \mathbf{5} & \mathbf{2} & \mathbf{1}
\end{array}
$$

We can even not perform any readjustment of the cba letter pattern over 31^3, in which case the calculation would be as follows:

$$
\begin{array}{rccccc}
 & c & b & a & & \\
\mathbf{31^3} = & \boxed{29} & \boxed{7} & \boxed{9} & 1 & \\
(\mathbf{3}-1)c = & 58 & & & & \\
(\mathbf{3}-1)b+\mathbf{1}c = & 4 & 3 & & & \\
(\mathbf{3}-1)a+\mathbf{1}b = & & 2 & 5 & & \\
+ & - & - & - & - & \\
 & \mathbf{92} & \mathbf{3} & \mathbf{4} & \mathbf{1} & \\
\boxed{ba}\,pattern\ (over\ 31^3) & * & * & \boxed{9} & \boxed{1} & \\
(\mathbf{3}-1)a+\mathbf{1}b = & & & \mathbf{1} & \mathbf{1} & \\
\mathbf{1}a = & & & & \mathbf{0} & \mathbf{1} \\
+ & - & - & - & - & - \\
\mathbf{31^4} = & 92 & 3 & 5 & 2 & 1
\end{array}
$$

With all this *we already have enough information to be able to solve the fourth root of a number.* We choose 25 $to\ the\ fourth\ power, plus$ 271 as the **radicand**, that is, 390 896. To find the first digit of the root, **we separate** *from right to left* and *four by four* the digits of the radicand **39 0896.** The number 2^4 doesn't exceed 39. The second digit of the root can be found as we did in the square root; then we only have to subtract the cube from the root and the addends that with it convert it into a power to the fourth:

$\sqrt[4]{3}$	9	$\underline{0}$	8	9	6	$\underline{\mathbf{2}}$ $\underline{\mathbf{5}}$ *Root*

Root **2**

$\mathbf{2}^4 = 16$

$39 - 16 = \mathbf{23}$

Root **5**

$2 \cdot \underline{\mathbf{2}} = \mathbf{4}$

$45 \cdot \underline{\mathbf{5}} = 225 < \mathbf{23}\underline{0}$

$46 \cdot 6 = 276 > \mathbf{23}\underline{0}$

−1	**−5**	**−6**	**−2**	**−5**
2	3	4	6	4

$\underline{\mathbf{2}}^2 = 4$ *

$2(\underline{2} \cdot \underline{\mathbf{5}}) = 2\ 0$ *

$\underline{\mathbf{5}}^2 = 2\ 5$ *

$(\underline{\mathbf{2}} - 1) \cdot \mathbf{6} = 6$

$(1) \cdot 2 + \underline{\mathbf{5}} \cdot 6 = 3\ 2$

$(1) \cdot 5 + \underline{\mathbf{5}} \cdot 2 = 1\ 5$

$\underline{\mathbf{5}} \cdot 5 = 2\ 5$

$\underline{\mathbf{25}}^3 += \mathbf{15}\ \mathbf{6}\ \bar{\bar{\bar{\mathbf{2}}}}\ \bar{\bar{\bar{\mathbf{5}}}}$

$\boxed{c}\ \boxed{b}\ \boxed{a}$

0	8			$(23 - \mathbf{15} = 8)$		$(\underline{\mathbf{2}} - 1) \cdot \boxed{15} = \mathbf{15}$
	0	3		$(84 - \mathbf{81} = 3)$		$(\underline{\mathbf{2}} - 1) \cdot b + \underline{\mathbf{5}} \cdot c$
$(36 - \mathbf{32} = 4)$			4			$(\underline{\mathbf{2}} - 1) \cdot a + \underline{\mathbf{5}} \cdot b$
$(44 - \mathbf{15} = 29)$			**2**	9		$(\underline{\mathbf{2}} - 1) \cdot \bar{\bar{\bar{5}}} + \underline{\mathbf{5}} \cdot \bar{\bar{\bar{2}}}$
$(96 - \mathbf{25} = 71)$				**7**	**1**	$\underline{\mathbf{5}}a = \underline{5} \cdot \bar{\bar{\bar{5}}}$
0	0	0	**2**	**7**	**1**	*(Remainder)*

The algorithm used **can be generalized** to solve for the *nth root of a number*. We only have to subtract from the radicand the number raised to a power less *(nth minus one)* and the addends that convert said power to the nth.

Converting a power to the next higher power is simple, although it depends on the digits of the number.

If the number has a single figure (u) one addend is enough for each digit of the lower power, that is, the product of said digit by one less ($u - 1$). For example, the *power* 8 (of 7) can be obtained from the *power* 7 (of 7) by adding to it the addends 48, 12, 18, 30, 24 and 18 *all properly shifted* from left to right in this way:

$$
\begin{array}{r|ccccccc}
\mathbf{7}^7 = & & 8 & 2 & 3 & 5 & 4 & 3 \\
(\mathbf{7}-1)\cdot 8 = & 4 & 8 & & & & & \\
(\mathbf{7}-1)\cdot 2 = & & 1 & 2 & & & & \\
(\mathbf{7}-1)\cdot 3 = & & & 1 & 8 & & & \\
(\mathbf{7}-1)\cdot 5 = & & & & 3 & 0 & & \\
(\mathbf{7}-1)\cdot 4 = & & & & & 2 & 4 & \\
(\mathbf{7}-1)\cdot 3 = & & & & & & 1 & 8 \\
+ & - & - & - & - & - & - & - \\
\mathbf{7}^8 = & \mathbf{5} & \mathbf{7} & \mathbf{6} & \mathbf{4} & \mathbf{8} & \mathbf{0} & \mathbf{1}
\end{array}
$$

If the number has 2 $digits$ $(DU = D \cdot 10 + U)$ a few addends are enough. With the $\boldsymbol{cba}$ **pattern** acting **on the leftmost digits** of the lower power:

$$\begin{array}{l}(\boldsymbol{D}-1)\cdot c \\ (\boldsymbol{D}-1)\cdot b+\boldsymbol{U}\cdot c \\ (\boldsymbol{D}-1)\cdot a+\boldsymbol{U}\cdot b\end{array}$$

With the $\boldsymbol{ba}$ **pattern** *shifted* **over** the digits of the lower power one position at a time until reaching the right end:

$$(\boldsymbol{D}-1)\cdot a+\boldsymbol{U}\cdot b$$

and finally, when the **pattern** $\boldsymbol{ba}$ is **over** the rightmost digits of the lower power:

$$\begin{array}{r}(\boldsymbol{D}-1)\cdot b+\boldsymbol{U}\cdot c \\ \boldsymbol{U}\cdot a\end{array}$$

never forgetting the shift to the right suffered by each of the addends except the first.

For example, to convert the $cube\ of$ 71 to its power to the fourth we add to the cube (357 911) the addends 18, 33, 47, 61, 15, 07 and 01, all except the first, *duly shifted* from *left to right* as follows:

$$
\begin{array}{r|cccccccccc}
\mathbf{71}^3 = & & 3 & 5 & 7 & 9 & 1 & 1 & \boldsymbol{c} & \boldsymbol{b} & \boldsymbol{a} \\
(\mathbf{7}-1)c = & \mathbf{1} & \mathbf{8} & & & & & & \boxed{3} & & \\
(\mathbf{7}-1)b+\mathbf{1}c = & & \mathbf{3} & \mathbf{3} & & & & & \boxed{3} & \boxed{5} & \\
(\mathbf{7}-1)a+\mathbf{1}b = & & & \mathbf{4} & \mathbf{7} & & & & & \boxed{5} & \boxed{7} \\
(\mathbf{7}-1)a+\mathbf{1}b = & & & & \mathbf{6} & \mathbf{1} & & & & \boxed{7} & \boxed{9} \\
(\mathbf{7}-1)a+\mathbf{1}b = & & & & & \mathbf{1} & \mathbf{5} & & & \boxed{9} & \boxed{1} \\
(\mathbf{7}-1)a+\mathbf{1}b = & & & & & & \mathbf{0} & \mathbf{7} & & \boxed{1} & \boxed{1} \\
\mathbf{1}a = & & & & & & & \mathbf{0} & \mathbf{1} & & \boxed{1} \\
+ & - & - & - & - & - & - & - & - & & \\
\mathbf{71}^4 = & 2 & 5 & 4 & 1 & 1 & 6 & 8 & 1 & &
\end{array}
$$

If the number has 3 $figures$ $(CDU = C \cdot 100 + D \cdot 10 + U)$ the pattern is still cba, but the addends include the hundreds of the number, although they follow the same logic when they are created. With the $\boldsymbol{cba}$ **pattern** acting **on the leftmost digits** of the lower power:

$$
\begin{array}{l}
(\boldsymbol{C}-1)\cdot c \\
(\boldsymbol{C}-1)\cdot b + \boldsymbol{D}\cdot c \\
(\boldsymbol{C}-1)\cdot a + \boldsymbol{D}\cdot b + \boldsymbol{U}\cdot c
\end{array}
$$

With the $\boldsymbol{cba}$ **pattern** shifted **over** the digits of the lower power one position at a time until reaching the right end:

$$(\boldsymbol{C}-1)\cdot a+\boldsymbol{D}\cdot b+\boldsymbol{U}\cdot c$$

and finally, when the ***ba* pattern** is **over** the rightmost digits of the lower power:

$$\boldsymbol{D}\cdot a+\boldsymbol{U}\cdot b$$
$$\boldsymbol{U}\cdot a$$

never forgetting the shift to the right suffered by each of the addends except the first:

$\mathbf{712^3} =$		3	6	0	9	4	4	1	2	8	$\boldsymbol{c}$	$\boldsymbol{b}$	$\boldsymbol{a}$
$(\mathbf{7}-1)c =$	1	8									$\boxed{3}$		
$(\mathbf{7}-1)b+\mathbf{1}c =$		3	9								$\boxed{3}$	$\boxed{6}$	
$(\mathbf{7}-1)a+\mathbf{1}b+\mathbf{2}c =$			1	2							$\boxed{3}$	$\boxed{6}$	$\boxed{0}$
$(\mathbf{7}-1)a+\mathbf{1}b+\mathbf{2}c =$				6	6						$\boxed{6}$	$\boxed{0}$	$\boxed{9}$
$(\mathbf{7}-1)a+\mathbf{1}b+\mathbf{2}c =$					3	3					$\boxed{0}$	$\boxed{9}$	$\boxed{4}$
$(\mathbf{7}-1)a+\mathbf{1}b+\mathbf{2}c =$						4	6				$\boxed{9}$	$\boxed{4}$	$\boxed{4}$
$(\mathbf{7}-1)a+\mathbf{1}b+\mathbf{2}c =$							1	8			$\boxed{4}$	$\boxed{4}$	$\boxed{1}$
$(\mathbf{7}-1)a+\mathbf{1}b+\mathbf{2}c =$								2	1		$\boxed{4}$	$\boxed{1}$	$\boxed{2}$
$(\mathbf{7}-1)a+\mathbf{1}b+\mathbf{2}c =$									5	2	$\boxed{1}$	$\boxed{2}$	$\boxed{8}$
$\mathbf{1}a+\mathbf{2}b =$										1	2	$\boxed{2}$	$\boxed{8}$
$\mathbf{2}a =$											1	6	$\boxed{8}$
$\mathbf{712^4} =$	**2**	**5**	**6**	**9**	**9**	**2**	**2**	**1**	**9**	**1**	**3**	**6**	

If the number has $4\,figures$ ($CDU = M\cdot 1000 + C\cdot 100 + D\cdot 10 + U$) the pattern is extended to $dcba$ and the addends include the thousands of the number, still following the same logic when they are created. With the $\boldsymbol{dcba}$ **pattern** acting **on the leftmost digits** of the lower power:

$$\begin{array}{l} (\boldsymbol{M}-1)\cdot d \\ (\boldsymbol{M}-1)\cdot b+\boldsymbol{C}\cdot c \\ (\boldsymbol{M}-1)\cdot a+\boldsymbol{C}\cdot b+\boldsymbol{D}\cdot c \\ (\boldsymbol{M}-1)\cdot a+\boldsymbol{C}\cdot b+\boldsymbol{D}\cdot c+\boldsymbol{U}\cdot d \end{array}$$

With the $\boldsymbol{dcba}$ **pattern** shifted **over** the digits of the lower power one position at a time until reaching the far right:

$$(\boldsymbol{M}-1)\cdot a+\boldsymbol{C}\cdot b+\boldsymbol{D}\cdot c+\boldsymbol{U}\cdot d$$

And finally, when the $\boldsymbol{cba}$ **pattern** is **over** the rightmost digits of the lower power:

$$\begin{array}{r} \boldsymbol{C}\cdot a+\boldsymbol{D}\cdot b+\boldsymbol{U}\cdot c \\ \boldsymbol{D}\cdot a+\boldsymbol{U}\cdot b \\ \boldsymbol{U}\cdot a \end{array}$$

never forgetting the shift to the right suffered by each of the addends except the first:

$\mathbf{3215}^2 =$	1	0	3	3	6	2	2	5	$\boldsymbol{d}$	$\boldsymbol{c}$	$\boldsymbol{b}$	$\boldsymbol{a}$
$(\mathbf{3}-1)d =$	2								$\boxed{1}$			
$(\mathbf{3}-1)c+\mathbf{2}d =$	0	2							$\boxed{1}$	$\boxed{0}$		
$(\mathbf{3}-1)b+\mathbf{2}c+\mathbf{1}d =$		0	7						$\boxed{1}$	$\boxed{0}$	$\boxed{3}$	
$(\mathbf{3}-1)a+\mathbf{2}b+\mathbf{1}c+\mathbf{5}d =$			1	7					$\boxed{1}$	$\boxed{0}$	$\boxed{3}$	$\boxed{3}$
$(\mathbf{3}-1)a+\mathbf{2}b+\mathbf{1}c+\mathbf{5}d =$				2	1				$\boxed{0}$	$\boxed{3}$	$\boxed{3}$	$\boxed{6}$
$(\mathbf{3}-1)a+\mathbf{2}b+\mathbf{1}c+\mathbf{5}d =$					3	4			$\boxed{3}$	$\boxed{3}$	$\boxed{6}$	$\boxed{2}$
$(\mathbf{3}-1)a+\mathbf{2}b+\mathbf{1}c+\mathbf{5}d =$						2	9		$\boxed{3}$	$\boxed{6}$	$\boxed{2}$	$\boxed{2}$
$(\mathbf{3}-1)a+\mathbf{2}b+\mathbf{1}c+\mathbf{5}d =$							4	6	$\boxed{6}$	$\boxed{2}$	$\boxed{2}$	$\boxed{5}$
$\mathbf{2}a+\mathbf{1}b+\mathbf{5}c =$								2	2	$\boxed{2}$	$\boxed{2}$	$\boxed{5}$
$\mathbf{1}a+\mathbf{5}b =$									1	5	$\boxed{2}$	$\boxed{5}$
$\mathbf{5}a =$										2	5	$\boxed{5}$
$\mathbf{3215}^3 =$	**3**	**3**	**2**	**3**	**0**	**9**	**6**	**3**	**3**	**7**	**5**	

With numbers with more figures, the dimension of the pattern is increased so that it covers all the digits of the number and proceeds as before, following the same logic, being careful with the necessary displacements.

It is also valid to use the cba pattern at all times, but in this case, we must be careful not to add the same thing twice. In the following example the cba pattern is fully used by applying it to the first numbers (321). In the final part, the pattern affects the last numbers (215) and the rest of the addends in which there is a collision, only the corresponding part is added:

$\mathbf{3215}^2 =$	1	0	3	3	6	2	2	5		c	b	a
$(\mathbf{3}-1)c =$	2									$\boxed{1}$		
$(\mathbf{3}-1)b+\mathbf{2}c =$	0	2								$\boxed{1}$	$\boxed{0}$	
$(\mathbf{3}-1)a+\mathbf{2}b+\mathbf{1}c =$		0	7							$\boxed{1}$	$\boxed{0}$	$\boxed{3}$
$(\mathbf{3}-1)a+\mathbf{2}b+\mathbf{1}c =$			1	2						$\boxed{0}$	$\boxed{3}$	$\boxed{3}$
$(\mathbf{3}-1)a+\mathbf{2}b+\mathbf{1}c =$				2	1					$\boxed{3}$	$\boxed{3}$	$\boxed{6}$
$(\mathbf{3}-1)a+\mathbf{2}b+\mathbf{1}c =$					1	9				$\boxed{3}$	$\boxed{6}$	$\boxed{2}$
$(\mathbf{3}-1)a+\mathbf{2}b+\mathbf{1}c =$						1	4			$\boxed{6}$	$\boxed{2}$	$\boxed{2}$
$(\mathbf{3}-1)a+\mathbf{2}b+\mathbf{1}c =$							1	6		$\boxed{2}$	$\boxed{2}$	$\boxed{5}$
$\mathbf{5}c =$			0	5						$\boxed{1}$	$\boxed{0}$	$\boxed{3}$
$\mathbf{5}c =$				0	0					$\boxed{0}$	$\boxed{3}$	$\boxed{3}$
$\mathbf{5}c =$					1	5				$\boxed{3}$	$\boxed{3}$	$\boxed{6}$
$\mathbf{5}c =$						1	5			$\boxed{3}$	$\boxed{6}$	$\boxed{2}$
$\mathbf{5}c =$							3	0		$\boxed{6}$	$\boxed{2}$	$\boxed{2}$
$\mathbf{2}a+\mathbf{1}b+\mathbf{5}c =$								2	2	$\boxed{2}$	$\boxed{2}$	$\boxed{5}$
$\mathbf{1}a+\mathbf{5}b =$									1	5	$\boxed{2}$	$\boxed{5}$
$\mathbf{5}a =$										2	5	$\boxed{5}$
$\mathbf{3215}^3 =$	**3**	**3**	**2**	**3**	**0**	**9**	**6**	**3**	**3**	**7**	**5**	

To solve the nth root of a number we only have to subtract from the radicand the number raised to a power less (nth minus one) and the addends that convert said power to the nth.

For each new digit of the root (of m-1 figures), we start from its last square to find the one of the current root (of m figures), and convert it to the lower power.

Chapter 9

Factoring a Number

A prime number can only be divided by itself and unity, but any other number is made up of a combination of primes multiplied together. In this chapter we will see how to find out what they are. To do this we will investigate when a number is divisible by the first natural numbers.

Division by 0

Cannot divide by 0; division by 0 is indeterminate.

Division by 1

All the numbers are divisible by 1.

Division by 2

A number is divisible by 2 when its last digit is 0, 2 or a multiple of 2.

Any number $N = a_n \cdots a_0 a_0$ can be put as a sum of powers of 10 as follows:

$$N = a_n \cdots a_0 a_0 = 10^n a_n + \cdots + 10^1 a_1 + 10^0 a_0$$

where a_0 is its last digit, a_n the first and 10^0 is 1.

Dividing N by 2 consists of dividing each of the terms that compose it by 2:

$$\frac{N}{2} = \frac{10^n a_n}{2} + \cdots + \frac{10^1 a_1}{2} + \frac{a_0}{2}$$

Since 10 divided by 2 is 5 and 10^n equals $10 \cdot 10^{n-1}$:

$$\frac{N}{2} = 10^{n-1} \cdot 5a_n + \cdots + 5a_1 + \frac{a_0}{2}$$

where $a_0/2$ is its last digit, which determines that the number will be divisible by 2 only when a_0 is, which is what we were looking for.

Division by 3

A number is divisible by 3 when the sum of its digits is 3 or a multiple of 3.

Dividing N by 3 consists of dividing each of the terms that compose it by 3:

$$\frac{N}{3} = \frac{10^n a_n}{3} + \cdots + \frac{10^1 a_1}{3} + \frac{a_0}{3}$$

By adding and subtracting each term $a_m/3$ $(m = 1 \ldots n)$ and grouping, we are left with the expression:

$$\frac{N}{3} = \left(\frac{10^n a_n}{3} - \frac{a_n}{3}\right) + \cdots + \left(\frac{10^1 a_1}{3} - \frac{a_1}{3}\right) + \frac{a_n}{3} + \cdots + \frac{a_0}{3}$$

We can easily check that the number before 10^m $(m = 1 \ldots n)$ is made up of m nines:

$$\begin{aligned}
10^1 - 1 = &\quad 10 - (10 - 9) &= 9 \\
10^2 - 1 = &\quad 100 - (100 - 99) &= 99 \\
\vdots &\quad \vdots &\vdots \\
10^m - 1 = &\quad 10\underbrace{\ldots}_{m}0 - \left(10\underbrace{\ldots}_{m}0 - 9\underbrace{\ldots}_{m}9\right) &= 9\underbrace{\ldots}_{m}9
\end{aligned}$$

then:

$$\frac{N}{3} = \left(\frac{10^n a_n - a_n}{3}\right) + \cdots + \left(\frac{10^1 a_1 - a_1}{3}\right) + \frac{a_n}{3} + \cdots + \frac{a_0}{3}$$

$$= \frac{(10^n - 1)a_n}{3} + \cdots + \frac{(9)a_1}{3} + \frac{a_n}{3} + \cdots + \frac{a_0}{3}$$

$$= \frac{\left(\overbrace{9 \ldots 9}^{n}\right) a_n}{3} + \cdots + \frac{(9)a_1}{3} + \frac{a_n}{3} + \cdots + \frac{a_0}{3}$$

$$= \left(\overbrace{3 \ldots 3}^{n}\right) a_n + \cdots + (3)a_1 + \frac{a_n + \cdots + a_0}{3}$$

which is precisely the result we were looking for, since the last term (the sum of the digits of the number divided by 3) determines that N will be divisible by 3 only if the sum $a_n + \cdots + a_0$ is divisible by 3.

Division by 4

A number is divisible by 4 when the number formed by its last two digits is 00, 04 or a multiple of 4.

Dividing N by 4 consists of dividing each of the terms that compose it by 4:

$$\frac{N}{4} = \frac{10^n a_n}{4} + \cdots + \frac{10^2 a_2}{4} + \frac{10^1 a_1}{4} + \frac{a_0}{4}$$

Since 100 *divided by* 4 is 25 and 10^n equals $100 \cdot 10^{n-2}$:

$$\frac{N}{4} = 25 \cdot 10^{n-2} a_n + \cdots + 25a_2 + \frac{10^1 a_1 + a_0}{4}$$

where $(10^1 a_1 + a_0)/4$ determines that the number is only divisible by 4 when $10^1 a_1 + a_0$ s, which is what we were looking for.

Division by 5

A number is divisible by 5 if its last digit is 0 or 5.

Dividing N by 5 consists of dividing each of the terms that compose it by 5:

$$\frac{N}{5} = \frac{10^n a_n}{5} + \cdots + \frac{10^1 a_1}{5} + \frac{a_0}{5}$$

Since 10 *divided by* 5 is 2 and 10^n equals $10 \cdot 10^{n-1}$:

$$\frac{N}{5} = 10^{n-1} \cdot 2a_n + \cdots + 2a_1 + \frac{a_0}{5}$$

where $a_0/5$ is its last digit, which determines that the number will be divisible by 5 only when a_0 is, which is what we were looking for.

Division by 6

A number is divisible by 6 when it is divisible by 2 and by 3 at the same time, that is, when its last digit is divisible by 2 and the sum of its digits is divisible by 3.

Division by 7

A number is divisible by 7 if the result of subtracting twice the units digit from said number without the units is 0, 7 or a multiple of 7.

Dividing a number a by another n produces a quotient k and a remainder b:

$$a = kn + b$$

an expression that ordered shows that $a - b$ is divisible by n, or equivalently, $a - b$ is a multiple of n:

$$\boldsymbol{a - b = kn}$$

and defines when a is congruent to b modulo n, which is expressed mathematically as:

$$\boldsymbol{a} \equiv \boldsymbol{b} \; (mod\; \boldsymbol{n}) \;\Leftrightarrow \exists\; \boldsymbol{k} \in \mathbb{Z} \;\; such\; that\; \boldsymbol{a - b = kn}$$

In addition, multiplying the above expression by any integer $\boldsymbol{m}$ or adding it to both sides of the equation maintains consistency:

$$\boldsymbol{A}) \quad a + \boldsymbol{m} \equiv b + \boldsymbol{m}\ (mod\ \boldsymbol{n})$$
$$\boldsymbol{B}) \quad a \cdot \boldsymbol{m} \equiv b \cdot \boldsymbol{m}\ (mod\ \boldsymbol{n})$$

Since $3 \cdot 7$ is exactly 21 we can say that when dividing 21 by 7 the remainder is 0, that is, 21 is 0 $(modulo\ \mathbf{7})$:

$$21 = 3 \cdot 7 + 0; \quad \mathbf{21 - 0 = 3 \cdot 7}; \quad \mathbf{21 \equiv 0}\ (mod\ \mathbf{7})$$

By property $\boldsymbol{B}$, multiplying both members by the integer a_0 maintains consistency:

$$21a_0 \equiv 0\ (mod\ 7)$$

Since 21 is 1 plus 20,

$$21a_0 = (1 + 20)a_0 = (\boldsymbol{a_0} + 20\boldsymbol{a_0}) \equiv \mathbf{0}\ (mod\ \mathbf{7})$$

and by property $\boldsymbol{A}$, adding $-20a_0$ to both members of the expression maintains consistency:

$$(a_0 + 20a_0) - 20a_0 = \boldsymbol{a_0} \equiv -20\boldsymbol{a_0}\ (mod\ \mathbf{7})$$

again, by property $\boldsymbol{A}$, we can add $10a_1$ to both sides of the equation:

$$10\boldsymbol{a_1} + \boldsymbol{a_0} \equiv 10\boldsymbol{a_1} - 20\boldsymbol{a_0}\ (mod\ \boldsymbol{7})$$

and since 20 is $10 \cdot 2$, taking out 10 common factor:

$$10\boldsymbol{a_1} + \boldsymbol{a_0} \equiv 10(\boldsymbol{a_1} - 2\boldsymbol{a_0})\ (mod\ \boldsymbol{7})$$

As the previous expression is general, it is possible to choose other integers a_0 and a_1; in particular, we can choose the same a_0 and replace $\boldsymbol{a_1}$ with $10\boldsymbol{a_2} + \boldsymbol{a_1}$:

$$\begin{aligned} 10(10\boldsymbol{a_2} + \boldsymbol{a_1}) + \boldsymbol{a_0} &\equiv 10(10\boldsymbol{a_2} + \boldsymbol{a_1} - 2\boldsymbol{a_0})\ (mod\ \boldsymbol{7}) \\ &\equiv 10^2\boldsymbol{a_2} + 10(\boldsymbol{a_1} - 2\boldsymbol{a_0})\ (mod\ \boldsymbol{7}) \end{aligned}$$

now we keep a_0 and a_1 unchanged and replace $\boldsymbol{a_2}$ with $10\boldsymbol{a_3} + \boldsymbol{a_2}$:

$$\begin{aligned} &10^2(10\boldsymbol{a_3} + \boldsymbol{a_2}) + 10\boldsymbol{a_1} + \boldsymbol{a_0} \\ &\qquad \equiv 10^2(10\boldsymbol{a_3} + \boldsymbol{a_2}) + 10(\boldsymbol{a_1} - 2\boldsymbol{a_0})\ (mod\ \boldsymbol{7}) \\ &\qquad \equiv 10^3\boldsymbol{a_3} + 10^2\boldsymbol{a_2} + 10(\boldsymbol{a_1} - 2\boldsymbol{a_0})\ (mod\ \boldsymbol{7}) \end{aligned}$$

Following the same logic, at each step we can leave all a_i intact (i from 0 to $n - 2$, both inclusive) and replace a_{n-1} with $10\boldsymbol{a_n} + \boldsymbol{a_{n-1}}$:

$$\begin{aligned}10^{n-1}(10\boldsymbol{a}_n + \boldsymbol{a}_{n-1}) + \cdots + \boldsymbol{a}_0 &\\ &\equiv 10^{n-1}(10\boldsymbol{a}_n + \boldsymbol{a}_{n-1}) + \cdots + 10(\boldsymbol{a}_1 - 2\boldsymbol{a}_0)\\ &\equiv 10^n\boldsymbol{a}_n + \cdots + 10(\boldsymbol{a}_1 - 2\boldsymbol{a}_0)\ (mod\ \mathbf{7})\\ &= (10^n\boldsymbol{a}_n + \cdots + 10\boldsymbol{a}_1) - 10(2\boldsymbol{a}_0)\ (mod\ \mathbf{7})\end{aligned}$$

that is, the number $N = (10^n\boldsymbol{a}_n + \cdots + \boldsymbol{a}_0)$ is equivalent, after being divided by 7, to N without units $(10^n\boldsymbol{a}_n + \cdots + 10\boldsymbol{a}_1)$ from which twice that term has been subtracted, just what we were looking for.

For example: 6251 is divisible by 7 if 623 (subtract $2 \cdot 1$ from 625) is and if 56 (subtract $2 \cdot 3$ from 62) is; but 56 is eight times seven (multiple of 7); therefore 6251 is divisible by 7.

Division by 8

A number is divisible by 8 when the number formed by its last three digits is 000, 008 or a multiple of 8.

Dividing N by 8 consists of dividing each of the terms that compose it by 8:

$$\frac{N}{8} = \frac{10^n a_n}{8} + \cdots + \frac{10^3 a_3}{8} + \frac{10^2 a_2}{8} + \frac{10^1 a_1}{8} + \frac{a_0}{8}$$

Since 1000 *divided by* 8 is 125 and 10^n is $1000 \cdot 10^{n-3}$:

$$\frac{N}{8} = 125 \cdot 10^{n-3} a_n + \cdots + 125 a_3 + \frac{10^2 a_2 + 10^1 a_1 + a_0}{8}$$

where $(10^2 a_2 + 10^1 a_1 + a_0)/8$ determines that the number is only divisible by 8 when $10^2 a_2 + 10^1 a_1 + a_0$ is, which is what we were looking for.

Division by 9

A number is divisible by 9 when the sum of its digits is 9 or a multiple of 9.

Dividing N by 9 consists of dividing each of the terms that compose it by 9:

$$\frac{N}{9} = \frac{10^n a_n}{9} + \cdots + \frac{10^1 a_1}{9} + \frac{a_0}{9}$$

By adding and subtracting each term $a_m/9$ $(m = 1 \ldots n)$ and grouping, we are left with the expression:

$$\frac{N}{9} = \left(\frac{10^n a_n}{9} - \frac{a_n}{9}\right) + \cdots + \left(\frac{10^1 a_1}{9} - \frac{a_1}{9}\right) + \frac{a_n}{9} + \cdots + \frac{a_0}{9}$$

Checking that the number just before 10^m $(m = 1 \ldots n)$ is made up of m nines is not complicated:

$$\begin{aligned}
10^1 - 1 &= \quad 10 - (10 - 9) &&= 9\\
10^2 - 1 &= \quad 100 - (100 - 99) &&= 99\\
&\vdots &&\vdots\\
10^m - 1 &= 10\underbrace{\ldots}_{m}0 - \left(10\underbrace{\ldots}_{m}0 - 9\underbrace{\ldots}_{m}9\right) &&= 9\underbrace{\ldots}_{m}9
\end{aligned}$$

applying this:

$$\begin{aligned}
\frac{N}{9} = \left(\frac{10^n a_n - a_n}{9}\right) + \cdots + \left(\frac{10^1 a_1 - a_1}{9}\right) + \frac{a_n}{9} + \cdots + \frac{a_0}{9}\\
= \frac{(10^n - 1)a_n}{9} + \cdots + \frac{(9)a_1}{9} + \frac{a_n}{9} + \cdots + \frac{a_0}{9}\\
= \frac{\left(9\overbrace{\ldots}^{n}9\right)a_n}{9} + \cdots + \frac{(9)a_1}{9} + \frac{a_n}{9} + \cdots + \frac{a_0}{9}\\
= \left(1\overbrace{\ldots}^{n}1\right)a_n + \cdots + (1)a_1 + \frac{a_n + \cdots + a_0}{9}
\end{aligned}$$

which is precisely the result we were looking for, since the last term (the sum of the digits of the number, divided by 9) determines that N will be divisible by 9 only if the sum $a_n + \cdots + a_0$ is divisible by 9.

Division by 10

A number is divisible by 10 if its last digit is 0.

Dividing N by 10 consists of dividing each of the terms that compose it by 10:

$$\frac{N}{10} = \frac{10^n a_n}{10} + \cdots + \frac{10^1 a_1}{10} + \frac{a_0}{10}$$

and operating,

$$\frac{N}{10} = 10^{n-1} a_n + \cdots + 10^1 a_2 + 10^0 a_1 + \frac{a_0}{10}$$

where $a_0/10$ determines that the number is only divisible by 10 when a_0 is. That implies that a_0 must be between 0 and 9 (both inclusive); but the only one divisible by 10 is 0 *(because 0 divided by any number is 0),* which is what we were looking for.

Division by 11

A number is divisible by 11 when the sum of its even place digits minus the sum of its odd place digits is 0, 11 or a multiple of 11.

Dividing N by 11 consists of dividing each of the terms that compose it by 11:

$$\frac{N}{11} = \frac{10^n a_n}{11} + \cdots + \frac{10^1 a_1}{11} + \frac{a_0}{11}$$

Since there is always a number $\boldsymbol{k}$ (generally different) that makes $\mathbf{10^n}$ equal to $11 \cdot \boldsymbol{k} + (-1)^{\boldsymbol{n}}$ for all $\boldsymbol{n}$ greater than or equal to zero (see proof at the end):

$$\frac{N}{11} = \frac{(11k + (-1)^n)a_n}{11} + \cdots + \frac{(11 \cdot 1 + (-1)^1)a_1}{11} + \frac{a_0}{11}$$

and rearranging addends,

$$\frac{N}{11} = \frac{11ka_n}{11} + \cdots + \frac{11a_1}{11} + \frac{(-1)^n a_n}{11} + \cdots + \frac{(-1)^1 a_1}{11} + \frac{a_0}{11}$$

Since $(-1)^n$ is 1 when n is even and (-1) when n is odd:

$$\frac{N}{11} = ka_n + \cdots + a_1 + \frac{(-1)^n a_n}{11} + \cdots - \frac{a_3}{11} + \frac{a_2}{11} - \frac{a_1}{11} + \frac{a_0}{11}$$

therefore, for N to be divisible by 11, the final terms must be divisible, that is, the sum of the digits of N of even place minus those of odd place (which is what we were looking for).

Lemma: *Always exists a number* $\mathbf{k}$ *(generally different) that makes* $\mathbf{10^n}$ *equal to* $11 \cdot \boldsymbol{k} + (-1)^{\boldsymbol{n}}$ *for all* $\mathbf{n}$ *greater than or equal to zero.*

Demonstration *(by method of induction on* n*):*

It's clear that the property holds for $n \leq 2$, since

$$\begin{aligned}
10^0 &= 1 = 0+1 = 11 \cdot 0 + (-1)^0 \\
10^1 &= 10 = 11-1 = 11 \cdot 1 + (-1)^1 \\
10^2 &= 100 = 99+1 = 11 \cdot 9 + (-1)^2
\end{aligned}$$

Suppose that it is still true for the first n integers (with $n \geq 2$), that is,

$$\mathbf{10^n} = 11 \cdot \boldsymbol{k} + (-1)^{\boldsymbol{n}}$$

Using the previous expression, we must deduce that this is true for the following:

$$10^{n+1} = \mathbf{10^n} \cdot 10 = (11 \cdot \boldsymbol{k} + (-1)^{\boldsymbol{n}}) \cdot (11-1)$$

operating and taking out 11 common factor:

$$\begin{aligned}
\mathbf{10^{n+1}} &= 11 \cdot 11 \cdot k + 11 \cdot (-1)^n + (-1) \cdot 11 \cdot k + (-1)^{n+1} \\
&= 11 \cdot (11k + (-1)^n + (-1)k) + (-1)^{n+1} \\
&= \mathbf{11k' + (-1)^{n+1}} q.e.d
\end{aligned}$$

Division by 12

A number is divisible by 12 when it is divisible by 3 and by 4 at the same time, that is, when the sum of its digits is divisible by 3 and the number formed by its last two digits is 00, 04 or multiple of 4.

Division by 13

A number is divisible by 13 if the result of subtracting nine times the units digit, from said number without the units, is 0, 13 or a multiple of 13.

We recall the definition of congruence:

$$\boldsymbol{a} \equiv \boldsymbol{b}\ (mod\ \boldsymbol{n}) \Leftrightarrow \exists\ \boldsymbol{k} \in \mathbb{Z}\ such\ that\ \boldsymbol{a} - \boldsymbol{b} = \boldsymbol{kn}$$

and that the congruence holds when multiplying by any integer $\boldsymbol{m}$ or adding it to both sides of the equation:

$$\begin{array}{ll} \boldsymbol{A}) & a + \boldsymbol{m} \equiv b + \boldsymbol{m}\ (mod\ \boldsymbol{n}) \\ \boldsymbol{B}) & a \cdot \boldsymbol{m} \equiv b \cdot \boldsymbol{m}\ (mod\ \boldsymbol{n}) \end{array}$$

Since $13 \cdot 7$ is exactly 91 we can say that when dividing 91 by 13 the remainder is 0, that is, 91 is $0\ (mod\ \mathbf{13})$:

$$91 = 13 \cdot 7 + 0; \quad \mathbf{91 - 0 = 13 \cdot 7}; \quad \mathbf{91 \equiv 0}\ (mod\ \mathbf{13})$$

By property $\boldsymbol{B}$, multiplying both members by the integer a_0 maintains consistency:

$$91a_0 \equiv 0\ (mod\ 7)$$

Since 91 equals 1 plus 90,

$$91a_0 = (1 + 90)a_0 = (\boldsymbol{a_0} + 90\boldsymbol{a_0}) \equiv \mathbf{0}\ (mod\ \mathbf{13})$$

and by property $\boldsymbol{A}$, adding $-90a_0$ to both members of the expression maintains consistency:

$$(a_0 + 90a_0) - 90a_0 = \boldsymbol{a_0} \equiv -90\boldsymbol{a_0}\ (mod\ \mathbf{13})$$

and again, by property $\boldsymbol{A}$, we can add $10a_1$ to both sides of the equation:

$$10\boldsymbol{a_1} + \boldsymbol{a_0} \equiv 10\boldsymbol{a_1} - 90\boldsymbol{a_0}\ (mod\ \mathbf{13})$$

and since 90 is $10 \cdot 9$, taking out 10 common factor:

$$10\boldsymbol{a_1} + \boldsymbol{a_0} \equiv 10(\boldsymbol{a_1} - 9\boldsymbol{a_0})\ (mod\ \mathbf{13})$$

As the previous expression is general, it's possible to choose other integers a_0 and a_1; in particular, we can choose the same a_0 and replace $\boldsymbol{a_1}$ with $10\boldsymbol{a_2} + \boldsymbol{a_1}$:

$$10(10\boldsymbol{a_2} + \boldsymbol{a_1}) + \boldsymbol{a_0} \equiv 10(10\boldsymbol{a_2} + \boldsymbol{a_1} - 9\boldsymbol{a_0})\ (mod\ \mathbf{13})$$
$$\equiv 10^2\boldsymbol{a_2} + 10(\boldsymbol{a_1} - 9\boldsymbol{a_0})\ (mod\ \mathbf{13})$$

now we keep a_0 and a_1 unchanged and replace $\boldsymbol{a_2}$ with

$10\boldsymbol{a_3} + \boldsymbol{a_2}$:

$$\begin{aligned}10^2(10\boldsymbol{a_3} + \boldsymbol{a_2}) + 10\boldsymbol{a_1} + \boldsymbol{a_0} \\ &\equiv 10^2(10\boldsymbol{a_3} + \boldsymbol{a_2}) + 10(\boldsymbol{a_1} - 9\boldsymbol{a_0})\ (mod\ \mathbf{13}) \\ &\equiv 10^3\boldsymbol{a_3} + 10^2\boldsymbol{a_2} + 10(\boldsymbol{a_1} - 9\boldsymbol{a_0})\ (mod\ \mathbf{13})\end{aligned}$$

Following the same logic, at each step we can leave all a_i intact (i from 0 to $n-2$, inclusive) and replace a_{n-1} with $10\boldsymbol{a_n} + \boldsymbol{a_{n-1}}$:

$$\begin{aligned}10^{n-1}(10\boldsymbol{a_n} + \boldsymbol{a_{n-1}}) + \cdots + \boldsymbol{a_0} \\ &\equiv 10^{n-1}(10\boldsymbol{a_n} + \boldsymbol{a_{n-1}}) + \cdots + 10(\boldsymbol{a_1} - 9\boldsymbol{a_0}) \\ &\equiv 10^n\boldsymbol{a_n} + \cdots + 10(\boldsymbol{a_1} - 9\boldsymbol{a_0})\ (mod\ \mathbf{13}) \\ &= (10^n\boldsymbol{a_n} + \cdots + 10\boldsymbol{a_1}) - 10 \cdot (9\boldsymbol{a_0})\ (mod\ \mathbf{13})\end{aligned}$$

that is, the number $N = (10^n\boldsymbol{a_n} + \cdots + \boldsymbol{a_0})$ is equivalent, after being divided by 13, to N without units $(10^n\boldsymbol{a_n} + \cdots + 10\boldsymbol{a_1})$ to which said term has been subtracted nine times, which is what we were looking for.

Example: 59 631 is divisible by 13 if 5954 (subtract $9 \cdot 1$ from 5963) is, and if 559 (subtract $9 \cdot 4$ from 595) is; but 559 equals 13 times 43 (that is, a multiple of 13); therefore, 59 631 is divisible by 13.

Division by a number N

A number is divisible by N if the result of adding the respective product of each of its figures (taken from right to left) by each digit of the repetitive series of remainders generated by dividing by N the different powers of 10 (from power 0 ad infinitum) is 0, N or multiple of N.

The remainders generated by dividing the powers of 10 by N, produce a series of numbers that repeat from a point to infinity or cyclically; for example, the sequence for **3** consists of **repeating the number 1 to infinity**:

$$\begin{aligned} 1 = 10^0 &= \mathbf{1} + 3 \cdot 0 \\ 10^1 &= \mathbf{1} + 3 \cdot 3 \\ 10^2 &= \mathbf{1} + 3 \cdot 33 \\ &\vdots \\ 10^n &= \mathbf{1} + 3 \cdot \overbrace{3 \dots 3}^{n} \end{aligned}$$

which coincides with the *criterion for dividing by 3,* namely that *the sum of all its figures must be divisible by 3 or by a multiple of 3:*

$$N = 10^n a_n + \cdots + a_0 = \mathbf{1}a_n + \cdots + \mathbf{1}a_0 \ (mod\ \mathbf{3})$$

The sequence for $\mathbf{6}$ is $\mathbf{1, 4, 4}$...:

$$\begin{aligned} 1 = 10^0 &= \mathbf{1} + 6 \cdot 0 \\ 10^1 &= \mathbf{4} + 6 \cdot 1 \\ 10^2 &= \mathbf{4} + 6 \cdot 16 \\ &\vdots \\ 10^n &= \mathbf{4} + 6 \cdot 16 \overbrace{\ldots}^{n-1} 6 \end{aligned}$$

that is,

$$N = 10^n a_n + \cdots + a_0 = \mathbf{4} a_n + \cdots + \mathbf{4} a_1 + \mathbf{1} a_0 \ (mod\ \mathbf{6})$$

or what is the same:

$$N = 10^n a_n + \cdots + a_0 = \mathbf{4}(a_n + \cdots + a_1) + a_0 \ (mod\ \mathbf{6})$$

which gives the following criterion: *A number is divisible by 6 if the sum of its last digit and the quadruple of said number without the units is 6 or a multiple of 6.*

For the number $\mathbf{7}$ the series of remainders is the repetition to infinity of the sequence $\mathbf{1, 3, 2, 6, 4, 5}$:

$$
\begin{aligned}
1 = 10^0 &= \boxed{1} + 7 \cdot 0 \\
10^1 &= \boxed{3} + 7 \cdot 1 \\
10^2 &= \boxed{2} + 7 \cdot 14 \\
10^3 &= \boxed{6} + 7 \cdot 142 \\
10^4 &= \boxed{4} + 7 \cdot 1428 \\
10^5 &= \boxed{5} + 7 \cdot 14285 \\
10^6 &= \mathbf{1} + 7 \cdot 142857 \\
10^7 &= \mathbf{3} + 7 \cdot 142857\mathbf{1} \\
10^8 &= \mathbf{2} + 7 \cdot 142857\mathbf{14} \\
10^9 &= \mathbf{6} + 7 \cdot 142857\mathbf{142} \\
&\vdots
\end{aligned}
$$

that is,

$$
\begin{aligned}
N = \; 10^n a_n + \cdots + a_7 + a_6 + a_5 + a_4 + a_3 + a_2 + a_1 + a_0 \\
= \cdots + \mathbf{3}a_7 + \mathbf{1}a_6 + \mathbf{5}a_5 + \mathbf{4}a_4 + \mathbf{6}a_3 + \mathbf{2}a_2 \\
+ \mathbf{3}a_1 + \mathbf{1}a_0 \; (mod\; \mathbf{7})
\end{aligned}
$$

which gives the following criterion: *A number is divisible by 7 if the result of adding the respective product of each of its figures (taken from right to left) by each digit of the series of remainders **1, 3, 2, 6, 4, 5**... is 7 or a multiple of 7.*

Example: 5376 is divisible by 7 since the result of the following sum is 63 ($multiple\ of$ 7):

$$\mathbf{6} \cdot 5 + \mathbf{2} \cdot 3 + \mathbf{3} \cdot 7 + \mathbf{1} \cdot 6 = 30 + 6 + 21 + 6 = 63 = 9 \cdot \mathbf{7}$$

The following table contains the series of remainders of the integers from **2** to **16**, where the number highlighted in a box indicates the place of beginning of the repetition of the sequence:

Remains	10^0	10^1	10^2	10^3	10^4	10^5	10^6	10^7	10^8
2	1	$\boxed{0}$	...						
3	$\boxed{1}$	...							
4	1	2	$\boxed{0}$	...					
5	1	$\boxed{0}$	...						
6	1	$\boxed{4}$	...						
7	$\boxed{1}$	3	2	6	4	5	...		
8	1	2	4	$\boxed{0}$	...				
9	$\boxed{1}$	...							
10	1	$\boxed{0}$	...						
11	$\boxed{1}$	10	...						
12	1	10	$\boxed{4}$	...					
13	$\boxed{1}$	10	9	12	3	4	...		
14	1	10	$\boxed{2}$	6	4	12	8	10	...
15	1	$\boxed{10}$	...						
16	1	10	4	8	$\boxed{0}$	...			

The following table is equivalent to the previous one, although the sum this time may be zero. E.g., 1111 is divisible by 11 since $\mathbf{1} \cdot 1 + (\mathbf{-1}) \cdot 1 + \mathbf{1} \cdot 1 + (\mathbf{-1}) \cdot 1 = 0$:

Remains	$\mathbf{10^0}$	$\mathbf{10^1}$	$\mathbf{10^2}$	$\mathbf{10^3}$	$\mathbf{10^4}$	$\mathbf{10^5}$	$\mathbf{10^6}$	$\mathbf{10^7}$	$\mathbf{10^8}$
2	1	$\boxed{0}$	...						
3	$\boxed{1}$	...							
4	1	2	$\boxed{0}$	...					
5	1	$\boxed{0}$	...						
6	1	$\boxed{4}$	...						
7	$\boxed{1}$	3	2	6	4	5	...		
8	1	2	4	$\boxed{0}$	...				
9	$\boxed{1}$	...							
10	1	$\boxed{0}$	...						
11	$\boxed{1}$	-1	...	$(10^1 = 11 \cdot 0 + \mathbf{10} = 11 \cdot 1 - \mathbf{1})$					
12	1	-2	$\boxed{4}$	...					
13	$\boxed{1}$	-3	9	-1	3	4	...		
14	1	-4	$\boxed{2}$	6	4	-2	8	-4	...
15	1	$\boxed{-5}$	...						
16	1	-6	4	8	$\boxed{0}$	...			

If necessary, the series of remainders of any other number can be calculated (objectively, it would be enough to draw criteria only for the primes) but it may not be practical because the series tend to be longer and more complex.

Prime numbers

A prime number is a natural number greater than 1 that is divisible by itself and by unity. We can easily get them as long as we start from the beginning by using the *Sieve of Eratosthenes;* knowing if a random number is prime is much more complex.

The **Sieve of Eratosthenes** consists of eliminating all the multiples of each objective number in order. All the remaining ones are prime numbers. The first prime is 2; we eliminate all its multiples, namely, 4 ($2 \; times \; 2$), 6 ($2 \; times \; 3$), 8 ($2 \; times \; 4$), etc.; then we proceed with 3 (which is prime because all the multiples of the primes less than it have been eliminated) repeating the same operation; now we eliminate all its multiples (those that have not yet been processed), namely, 9 (3 *times* 3), 15 (3 *times* 5), etc., and so on until we reach the proposed goal (since the primes are infinite).

The primes less than or equal to 25 obtained using the Sieve of Eratosthenes are the 9 boxed:

	$\boxed{2}$	$\boxed{3}$	2^2	$\boxed{5}$
$2;3$	$\boxed{7}$	2^3	3^2	$2;5$
$\boxed{11}$	$2^2;3$	$\boxed{13}$	$2;7$	$3;5$
2^4	$\boxed{17}$	$2;3^2$	$\boxed{19}$	$2^2;5$
$3;7$	$2;11$	$\boxed{23}$	$2^3;3$	5^2

The primes between 26 and 50 are the 6 boxed:

$2;13$	3^3	$2^2;7$	$\boxed{29}$	$2;3;5$
$\boxed{31}$	2^5	$3;11$	$2;17$	$5;7$
$2^2;3^2$	$\boxed{37}$	$2;19$	$3;13$	$2^3;5$
$\boxed{41}$	$2;3;7$	$\boxed{43}$	$2^2;11$	$3^2;5$
$2;23$	$\boxed{47}$	$2^4;3$	7^2	$2;5^2$

The primes between 51 and 75 are the 6 boxed:

$3;17$	$2^2;13$	$\boxed{53}$	$2;3^3$	$5;11$
$2^3;7$	$3;19$	$2;29$	$\boxed{59}$	$2^2;3;5$
$\boxed{61}$	$2;31$	$3^2;7$	2^6	$5;13$
$2;3;11$	$\boxed{67}$	$2^2;17$	$3;23$	$2;5;7$
$\boxed{71}$	$2^3;3^2$	$\boxed{73}$	$2;37$	$3;5^2$

The primes between 76 and 100 are the 4 boxed:

$2^2; 19$	$7; 11$	$2; 3; 13$	$\boxed{79}$	$2^4; 5$
3^4	$2; 41$	$\boxed{83}$	$2^2; 3; 7$	$5; 17$
$2; 43$	$3; 29$	$2^3; 11$	$\boxed{89}$	$2; 3^2; 5$
$7; 13$	$2^2; 23$	$3; 31$	$2; 47$	$5; 19$
$2^5; 3$	$\boxed{97}$	$2; 7^2$	$3^2; 11$	$2^2; 5^2$

The primes between 101 and 125 are the 5 boxed:

$\boxed{101}$	$2; 3; 17$	$\boxed{103}$	$2^3; 13$	$3; 5; 7$
$2; 53$	$\boxed{107}$	$2^2; 3^3$	$\boxed{109}$	$2; 5; 11$
$3; 37$	$2^4; 7$	$\boxed{113}$	$2; 3; 19$	$5; 23$
$2^2; 29$	$3^2; 13$	$2; 59$	$7; 17$	$2^3; 3; 5$
11^2	$2; 61$	$3; 41$	$2^2; 31$	5^3

The primes between 126 and 150 are the 5 boxed:

$2; 3^2; 7$	$\boxed{127}$	2^7	$3; 43$	$2; 5; 13$
$\boxed{131}$	$2^2; 3; 11$	$7; 19$	$2; 67$	$3^3; 5$
$2^3; 17$	$\boxed{137}$	$2; 3; 23$	$\boxed{139}$	$2^2; 5; 7$
$3; 47$	$2; 71$	$11; 13$	$2^4; 3^2$	$5; 29$
$2; 73$	$3; 7^2$	$2^2; 37$	$\boxed{149}$	$2; 3; 5^2$

The primes between 151 and 175 are the 5 boxed:

$\boxed{151}$	$2^3; 19$	$3^2; 17$	$2; 7; 11$	$5; 31$
$2^2; 3; 13$	$\boxed{157}$	$2; 79$	$3; 53$	$2^5; 5$
$7; 23$	$2; 3^4$	$\boxed{163}$	$2^2; 41$	$3; 5; 11$
$2; 83$	$\boxed{167}$	$2^3; 3; 7$	13^2	$2; 5; 17$
$3^2; 19$	$2^2; 43$	$\boxed{173}$	$2; 3; 29$	$5^2; 7$

The primes between 176 and 200 are the 6 boxed:

$2^4; 11$	$3; 59$	$2; 89$	$\boxed{179}$	$2^2; 3^2; 5$
$\boxed{181}$	$2; 7; 13$	$3; 61$	$2^3; 23$	$5; 37$
$2; 3; 31$	$11; 17$	$2^2; 47$	$3^3; 7$	$2; 5; 19$
$\boxed{191}$	$2^6; 3$	$\boxed{193}$	$2; 97$	$3; 5; 13$
$2^2; 7^2$	$\boxed{197}$	$2; 9; 11$	$\boxed{199}$	$2^3; 5^2$

The primes between 201 and 225 are the 2 boxed:

$3; 67$	$2; 101$	$7; 29$	$2^2; 3; 17$	$5; 41$
$2; 103$	$3^2; 23$	$2^4; 13$	$11; 19$	$2; 3; 5; 7$
$\boxed{211}$	$2^2; 53$	$3; 71$	$2; 107$	$5; 43$
$2^3; 3^3$	$7; 31$	$2; 109$	$3; 73$	$2^2; 5; 11$
$13; 17$	$2; 3; 37$	$\boxed{223}$	$2^5; 7$	$3^2; 5^2$

The primes between 226 and 250 are the 5 boxed:

$2; 113$	$\boxed{227}$	$2^2; 3; 19$	$\boxed{229}$	$2; 5; 23$
$3; 7; 11$	$2^3; 29$	$\boxed{233}$	$2; 3^2; 13$	$5; 47$
$2^2; 59$	$3; 79$	$2; 7; 17$	$\boxed{239}$	$2^4; 3; 5$
$\boxed{241}$	$2; 11^2$	3^5	$4; 61$	$5; 7^2$
$2; 3; 41$	$13; 19$	$2^3; 31$	$3; 83$	$2; 5^3$

The primes between 251 and 275 are the 5 boxed:

$\boxed{251}$	$2^2; 3^2; 7$	$11; 23$	$2; 127$	$3; 5; 17$
2^8	$\boxed{257}$	$2; 3; 43$	$7; 37$	$2^2; 5; 13$
$3^2; 29$	$2; 131$	$\boxed{263}$	$2^3; 3; 11$	$5; 53$
$2; 7; 19$	$3; 89$	$2^2; 67$	$\boxed{269}$	$2; 3^3; 5$
$\boxed{271}$	$2^4; 17$	$3; 7; 13$	$2; 137$	$5^2; 11$

The primes between 276 and 300 are the 4 boxed:

$2^2; 3; 23$	$\boxed{277}$	$2; 139$	$3^2; 31$	$2^3; 5; 7$
$\boxed{281}$	$2; 3; 47$	$\boxed{283}$	$2^2; 71$	$3; 5; 19$
$2; 11; 13$	$7; 41$	$2^5; 3^2$	17^2	$2; 5; 29$
$3; 97$	$2^2; 73$	$\boxed{293}$	$2; 3; 7^2$	$5; 59$
$2^3; 37$	$3^3; 11$	$2; 149$	$13; 23$	$2^2; 3; 5^2$

The primes between 301 and 325 are the 4 boxed:

$7; 43$	$2; 151$	$3; 101$	$2^4; 19$	$5; 61$
$2; 3^2; 17$	$\boxed{307}$	$2^2; 7; 11$	$3; 103$	$2; 5; 31$
$\boxed{311}$	$2^3; 3; 13$	$\boxed{313}$	$2; 157$	$3^2; 5; 7$
$2^2; 79$	$\boxed{317}$	$2; 3; 53$	$11; 29$	$2^6; 5$
$3; 107$	$2; 7; 23$	$17; 19$	$2^2; 3^4$	$5^2; 13$

The primes between 326 and 350 are the 4 boxed:

$2; 163$	$3; 109$	$2^3; 41$	$7; 47$	$2; 3; 5; 11$
$\boxed{331}$	$2^2; 83$	$3^2; 37$	$2; 167$	$5; 67$
$2^4; 3; 7$	$\boxed{337}$	$2; 13^2$	$3; 113$	$2^2; 5; 17$
$11; 31$	$2; 3^2; 19$	7^3	$2^3; 43$	$3; 5; 23$
$2; 173$	$\boxed{347}$	$2^2; 3; 29$	$\boxed{349}$	$2; 5^2; 7$

The primes between 351 and 375 are the 4 boxed:

$3^3; 13$	$2^5; 11$	$\boxed{353}$	$2; 3; 59$	$5; 71$
$2^2; 89$	$3; 7; 17$	$2; 179$	$\boxed{359}$	$2^3; 3^2; 5$
19^2	$2; 181$	$3; 11^2$	$2^2; 7; 13$	$5; 73$
$2; 3; 61$	$\boxed{367}$	$2^4; 23$	$9; 41$	$2; 5; 37$
$7; 53$	$2^2; 3; 31$	$\boxed{373}$	$2; 11; 17$	$3; 5^3$

The primes between 376 and 400 are the 4 boxed:

$2^3; 47$	$13; 29$	$2; 3^3; 7$	$\boxed{379}$	$2^2; 5; 19$
$3; 127$	$2; 191$	$\boxed{383}$	$2^7; 3$	$5; 7; 11$
$2; 193$	$3^2; 43$	$2^2; 97$	$\boxed{389}$	$2; 3; 5; 13$
$17; 23$	$2^3; 7^2$	$3; 131$	$2; 197$	$5; 79$
$2^2; 3^2; 11$	$\boxed{397}$	$2; 199$	$3; 7; 19$	$2^4; 5^2$

The primes between 401 and 425 are the 4 boxed:

$\boxed{401}$	$2; 3; 67$	$13; 31$	$4; 101$	$3^4; 5$
$2; 7; 29$	$11; 37$	$2^3; 3; 17$	$\boxed{409}$	$2; 5; 41$
$3; 137$	$2^2; 103$	$7; 59$	$2; 3^2; 23$	$5; 83$
$2^5; 13$	$3; 139$	$2; 11; 19$	$\boxed{419}$	$2^2; 3; 5; 7$
$\boxed{421}$	$2; 211$	$3^2; 47$	$2^3; 53$	$5^2; 17$

The primes between 426 and 450 are the 5 boxed:

$2; 3; 71$	$7; 61$	$2^2; 107$	$3; 11; 13$	$2; 5; 43$
$\boxed{431}$	$2^4; 3^3$	$\boxed{433}$	$2; 7; 31$	$3; 5; 29$
$2^2; 109$	$19; 23$	$2; 3; 73$	$\boxed{439}$	$2^3; 5; 11$
$3^2; 7^2$	$2; 13; 17$	$\boxed{443}$	$2^2; 3; 37$	$5; 89$
$2; 223$	$3; 149$	$2^6; 7$	$\boxed{449}$	$2; 3^2; 5^2$

The primes between 451 and 475 are the 4 boxed:

$11; 41$	$2^2; 113$	$3; 151$	$2; 227$	$5; 7; 13$
$2^3; 3; 19$	$\boxed{457}$	$2; 229$	$3^3; 17$	$2^2; 5; 23$
$\boxed{461}$	$2; 3; 7; 11$	$\boxed{463}$	$2^4; 29$	$3; 5; 31$
$2; 233$	$\boxed{467}$	$2^2; 3^2; 13$	$7; 67$	$2; 5; 47$
$3; 157$	$2^3; 59$	$11; 43$	$2; 3; 79$	$5^2; 19$

The primes between 476 and 500 are the 4 boxed:

$2^2; 7; 17$	$3^2; 53$	$2; 239$	$\boxed{479}$	$2^5; 3; 5$
$13; 37$	$2; 241$	$3; 7; 23$	$2^2; 11^2$	$5; 97$
$2; 3^5$	$\boxed{487}$	$2^3; 61$	$3; 163$	$2; 5; 7^2$
$\boxed{491}$	$2^2; 3; 41$	$17; 29$	$2; 13; 19$	$3^2; 5; 11$
$2^4; 31$	$7; 71$	$2; 3; 83$	$\boxed{499}$	$2^2; 5^3$

The primes between 501 and 525 are the 4 boxed:

$3; 167$	$2; 251$	$\boxed{503}$	$2^3; 3^2; 7$	$5; 101$
$2; 11; 23$	$3; 13^2$	$2^2; 127$	$\boxed{509}$	$2; 3; 5; 17$
$7; 73$	2^9	$3^3; 19$	$2; 257$	$5; 103$
$2^2; 3; 43$	$11; 47$	$2; 7; 37$	$3; 173$	$2^3; 5; 13$
$\boxed{521}$	$2; 3^2; 29$	$\boxed{523}$	$2^2; 131$	$3; 5^2; 7$

The primes between 526 and 550 are the 2 boxed:

$2; 263$	$17; 31$	$2^4; 3; 11$	23^2	$2; 5; 53$
$3^2; 59$	$2^2; 7; 19$	$13; 41$	$2; 3; 89$	$5; 107$
$2^3; 67$	$3; 179$	$2; 269$	$7^2; 11$	$2^2; 3^3; 5$
$\boxed{541}$	$2; 271$	$3; 181$	$2^5; 17$	$5; 109$
$2; 3; 7; 13$	$\boxed{547}$	$2^2; 137$	$3^2; 61$	$2; 5^2; 11$

The primes between 551 and 575 are the 4 boxed:

$19; 29$	$2^3; 3; 23$	$7; 79$	$2; 277$	$3; 5; 37$
$2^2; 139$	$\boxed{557}$	$2; 3^2; 31$	$13; 43$	$2^4; 5; 7$
$3; 11; 17$	$2; 281$	$\boxed{563}$	$2^2; 3; 47$	$5; 113$
$2; 283$	$3^4; 7$	$2^3; 71$	$\boxed{569}$	$2; 3; 5; 19$
$\boxed{571}$	$2^2; 11; 13$	$3; 191$	$2; 7; 41$	$5^2; 23$

The primes between 576 and 600 are the 4 boxed:

$2^6;3^2$	$\boxed{577}$	$2;17^2$	$3;193$	$2^2;5;29$
$7;83$	$2;3;97$	$11;53$	$2^3;73$	$3^2;5;13$
$2;293$	$\boxed{587}$	$2^2;3;7^2$	$19;31$	$2;5;59$
$3;197$	$2^4;37$	$\boxed{593}$	$2;3^3;11$	$5;7;17$
$2^2;149$	$3;199$	$2;13;23$	$\boxed{599}$	$2^3;3;5^2$

The primes between 601 and 625 are the 5 boxed:

$\boxed{601}$	$2;7;43$	$3^2;67$	$2^2;151$	$5;11^2$
$2;3;101$	$\boxed{607}$	$2^5;19$	$3;7;29$	$2;5;61$
$13;47$	$2^2;3^2;17$	$\boxed{613}$	$2;307$	$3;5;41$
$2^3;7;11$	$\boxed{617}$	$2;3;103$	$\boxed{619}$	$2^2;5;31$
$3^3;23$	$2;311$	$7;89$	$2^4;3;13$	5^4

The primes between 626 and 650 are the 4 boxed:

$2;313$	$3;11;19$	$2^2;157$	$17;37$	$2;3^2;5;7$
$\boxed{631}$	$2^3;79$	$3;211$	$2;317$	$5;127$
$2^2;3;53$	$7^2;13$	$2;11;29$	$3^2;71$	$2^7;5$
$\boxed{641}$	$2;3;107$	$\boxed{643}$	$2^2;7;23$	$3;5;43$
$2;17;19$	$\boxed{647}$	$2^3;3^4$	$11;59$	$2;5^2;13$

The primes between 651 and 675 are the 4 boxed:

$3;7;31$	$2^2;163$	$\boxed{653}$	$2;3;109$	$5;131$
$2^4;41$	$3^2;73$	$2;7;47$	$\boxed{659}$	$2^2;3;5;11$
$\boxed{661}$	$2;331$	$3;13;17$	$2^3;83$	$5;7;19$
$2;3^2;37$	$23;29$	$2^2;167$	$3;223$	$2;5;67$
$11;61$	$2^5;3;7$	$\boxed{673}$	$2;337$	$3^3;5^2$

The primes between 676 and 700 are the 3 boxed:

$2^2; 13^2$	$\boxed{677}$	$2; 3; 113$	$7; 97$	$2^3; 5; 17$
$3; 227$	$2; 11; 31$	$\boxed{683}$	$2^2; 3^2; 19$	$5; 137$
$2; 7^3$	$3; 229$	$2^4; 43$	$13; 53$	$2; 3; 5; 23$
$\boxed{691}$	$2^2; 173$	$3^2; 7; 11$	$2; 347$	$5; 139$
$2^3; 3; 29$	$17; 41$	$2; 349$	$3; 233$	$2^2; 5^2; 7$

The primes between 701 and 725 are the 3 boxed:

$\boxed{701}$	$2; 3^3; 13$	$19; 37$	$2^6; 11$	$3; 5; 47$
$2; 353$	$7; 101$	$2^2; 3; 59$	$\boxed{709}$	$2; 5; 71$
$3^2; 79$	$2^3; 89$	$23; 31$	$2; 3; 7; 17$	$5; 11; 13$
$2^2; 179$	$3; 239$	$2; 359$	$\boxed{719}$	$2^4; 3^2; 5$
$7; 103$	$2; 19^2$	$3; 241$	$2^2; 181$	$5^2; 29$

The primes between 726 and 750 are the 4 boxed:

$6; 11^2$	$\boxed{727}$	$2^3; 7; 13$	3^6	$2; 5; 73$
$17; 43$	$2^2; 3; 61$	$\boxed{733}$	$2; 367$	$3; 5; 7^2$
$2^5; 23$	$11; 67$	$2; 3^2; 41$	$\boxed{739}$	$2^2; 5; 37$
$3; 13; 19$	$2; 7; 53$	$\boxed{743}$	$2^3; 3; 31$	$5; 149$
$2; 373$	$3^2; 83$	$2^2; 11; 17$	$7; 107$	$2; 3; 5^3$

The primes between 751 and 775 are the 5 boxed:

$\boxed{751}$	$2^4; 47$	$3; 251$	$2; 13; 29$	$5; 151$
$2^2; 3^3; 7$	$\boxed{757}$	$2; 379$	$3; 11; 23$	$2^3; 5; 19$
$\boxed{761}$	$2; 3; 127$	$7; 109$	$2^2; 191$	$3^2; 5; 17$
$2; 383$	$13; 59$	$2^8; 3$	$\boxed{769}$	$2; 5; 7; 11$
$3; 257$	$2^2; 193$	$\boxed{773}$	$2; 3^2; 43$	$5^2; 31$

The primes between 776 and 800 are the 2 boxed:

$8; 97$	$3; 7; 37$	$2; 389$	$19; 41$	$2^2; 3; 5; 13$
$11; 71$	$2; 17; 23$	$3^3; 29$	$2^4; 7^2$	$5; 157$
$2; 3; 131$	$\boxed{787}$	$2^2; 197$	$3; 263$	$2; 5; 79$
$7; 113$	$2^3; 3^2; 11$	$13; 61$	$2; 397$	$3; 5; 53$
$2^2; 199$	$\boxed{797}$	$2; 3; 7; 19$	$17; 47$	$2^5; 5^2$

The primes between 801 and 825 are the 4 boxed:

$3^2; 89$	$2; 401$	$11; 73$	$2^2; 3; 67$	$5; 7; 23$
$2; 13; 31$	$3; 269$	$2^3; 101$	$\boxed{809}$	$2; 3^4; 5$
$\boxed{811}$	$2^2; 7; 29$	$3; 271$	$2; 11; 37$	$5; 163$
$2^4; 3; 17$	$19; 43$	$2; 409$	$3^2; 7; 13$	$2^2; 5; 41$
$\boxed{821}$	$2; 3; 137$	$\boxed{823}$	$2^3; 103$	$3; 5^2; 11$

The primes between 826 and 850 are the 3 boxed:

$2; 7; 59$	$\boxed{827}$	$2^2; 3^2; 23$	$\boxed{829}$	$2; 5; 83$
$3; 277$	$2^6; 13$	$7^2; 17$	$2; 3; 139$	$5; 167$
$4; 11; 19$	$27; 31$	$2; 419$	$\boxed{839}$	$2^3; 3; 5; 7$
$29; 29$	$2; 421$	$3; 281$	$2^2; 211$	$5; 13^2$
$2; 3^2; 47$	$7; 11^2$	$2^4; 53$	$3; 283$	$2; 5^2; 17$

The primes between 851 and 875 are the 4 boxed:

$23; 37$	$2^2; 3; 71$	$\boxed{853}$	$2; 7; 61$	$3^2; 5; 19$
$2^3; 107$	$\boxed{857}$	$2; 3; 11; 13$	$\boxed{859}$	$2^2; 5; 43$
$3; 7; 41$	$2; 431$	$\boxed{863}$	$2^5; 3^3$	$5; 173$
$2; 433$	$3; 17^2$	$2^2; 7; 31$	$11; 79$	$2; 3; 5; 29$
$13; 67$	$2^3; 109$	$3^2; 97$	$2; 19; 23$	$5^3; 7$

The primes between 876 and 900 are the 4 boxed:

$2^2; 3; 73$	$\boxed{877}$	$2; 439$	$3; 293$	$2^4; 5; 11$
$\boxed{881}$	$2; 3^2; 7^2$	$\boxed{883}$	$2^2; 13; 17$	$3; 5; 59$
$2; 443$	$\boxed{887}$	$2^3; 3; 37$	$7; 127$	$2; 5; 89$
$3^4; 11$	$2^2; 223$	$19; 47$	$2; 3; 149$	$5; 179$
$2^7; 7$	$3; 13; 23$	$2; 449$	$29; 31$	$2^2; 3^2; 5^2$

The primes between 901 and 925 are the 3 boxed:

$17; 53$	$2; 11; 41$	$3; 7; 43$	$2^3; 113$	$5; 181$
$2; 3; 151$	$\boxed{907}$	$2^2; 227$	$3^2; 101$	$2; 5; 7; 13$
$\boxed{911}$	$2^4; 3; 19$	$11; 83$	$2; 457$	$3; 5; 61$
$2^2; 229$	$7; 131$	$2; 3^3; 17$	$\boxed{919}$	$2^3; 5; 23$
$3; 307$	$2; 461$	$13; 71$	$2^2; 3; 7; 11$	$5^2; 37$

The primes between 926 and 950 are the 4 boxed:

$2; 463$	$3^2; 103$	$2^5; 29$	$\boxed{929}$	$2; 3; 5; 31$
$7^2; 19$	$2^2; 233$	$3; 311$	$2; 467$	$5; 11; 17$
$2^3; 3^2; 13$	$\boxed{937}$	$2; 7; 67$	$3; 313$	$2^2; 5; 47$
$\boxed{941}$	$2; 3; 157$	$23; 41$	$2^4; 59$	$3^3; 5; 7$
$2; 11; 43$	$\boxed{947}$	$2^2; 3; 79$	$13; 73$	$2; 5^2; 19$

The primes between 951 and 975 are the 3 boxed:

$3; 317$	$2^3; 7; 17$	$\boxed{953}$	$2; 3^2; 53$	$5; 191$
$2^2; 239$	$3; 11; 29$	$2; 479$	$7; 137$	$2^6; 3; 5$
31^2	$2; 13; 37$	$3^2; 107$	$2^2; 241$	$5; 193$
$2; 3; 7; 23$	$\boxed{967}$	$2^3; 11^2$	$3; 17; 19$	$2; 5; 97$
$\boxed{971}$	$2^2; 3^5$	$7; 139$	$2; 487$	$3; 5^2; 13$

The primes between 976 and 1000 are the 4 boxed:

$2^4; 61$	$\boxed{977}$	$2; 3; 163$	$11; 89$	$2^2; 5; 7^2$
$3^2; 109$	$2; 491$	$\boxed{983}$	$2^3; 3; 41$	$5; 197$
$2; 17; 29$	$3; 7; 47$	$2^2; 13; 19$	$23; 43$	$2; 3^2; 5; 11$
$\boxed{991}$	$2^5; 31$	$3; 331$	$2; 7; 71$	$5; 199$
$2^2; 3; 83$	$\boxed{997}$	$2; 499$	$27; 37$	$2^3; 5^3$

Factorial decomposition of a number

Given any number, it is either prime or it's made up of a combination of primes multiplied by each other. The classic way to find out looks like this:

D	*i*	*v*	*i*	*d*	*e*	*n*	*d*		*Divisor (prime)*
3	**5**	**9**	**4**	**5**	**9**	**1**	**0**	**0**	5
	7	1	8	9	1	8	2	0	5
	1	4	3	7	8	3	6	4	2
		7	1	8	9	1	8	2	2
		3	5	9	4	5	9	1	3
		1	1	9	8	1	9	7	3
			3	9	9	3	9	9	3
			1	3	3	1	3	3	11
				1	2	1	0	3	7
					1	7	2	9	7
						2	4	7	13
							1	9	19
								1	1

whence it follows that:

$$359\,459\,100 = 1 \cdot 2^2 \cdot 3^3 \cdot 5^2 \cdot 7^2 \cdot 11 \cdot 13 \cdot 19$$

It is usual to use mostly the criterion of 2, 3, 5, 7, 11 and 13 even if the number is divisible by one of its multiples. *The algorithm is simple; we only need to divide the dividend by the corresponding prime and replace it with the generated quotient until it is 1.* It is possible to reduce calculations when dividing by the multiple of a known prime. Using the example above:

D	*i*	*v*	*i*	*d*	*e*	*n*	*d*		*Divisor*
3	**5**	**9**	**4**	**5**	**9**	**1**	**0**	**0**	$10\ (2 \cdot 5)$
	3	5	9	4	5	9	1	0	$10\ (2 \cdot 5)$
		3	5	9	4	5	9	1	$9\ (3^2)$
			3	9	9	3	9	9	3
			1	3	3	1	3	3	11
				1	2	1	0	3	7
					1	7	2	9	7
						2	4	7	13
							1	9	19
								1	1

In the example, 359 459 100 ends in 0 so it can be divided by 10; 35 945 910 ends in 0 so it can also be divided by 10; adding the figures of 3 594 591 we obtain 36 (*multiple of* 9) therefore, it is divisible by 9; now the dividend is the last generated quotient, that is, 399 399 which is clearly a multiple of 3; the generated dividend is **133 133** whose structure is typical of a multiple of 11 (even and odd numbers add up to the same thing, so their difference is zero); the new dividend is 12 103 which is divisible by 7 — since 1210 minus twice 3 is 1204; 120 minus 8 (*twice* 4) is 112; y 11 minus 4 (*twice* 2) is 7—; the corresponding quotient (1729) is too (because 172 minus 18 is 154; and 15 minus 8 is 7); the new dividend (247) is divisible by 13, since 24 minus $9 \cdot 7$ (= 63) equals −39, which is triple 13; finally, 19 is *prime* and is only divisible by itself. In this case we get:

$$\begin{aligned}\mathbf{359\,459\,100} &= 1 \cdot 3 \cdot 7^2 \cdot 9 \cdot 10^2 \cdot 11 \cdot 13 \cdot 19 \\ &= 1 \cdot 3 \cdot 7^2 \cdot 3^2 \cdot (2 \cdot 5)^2 \cdot 11 \cdot 13 \cdot 19 \\ &= 1 \cdot 3 \cdot 7^2 \cdot 3^2 \cdot 2^2 \cdot 5^2 \cdot 11 \cdot 13 \cdot 19 \\ &= 1 \cdot 2^2 \cdot 3 \cdot 3^2 \cdot 5^2 \cdot 7^2 \cdot 11 \cdot 13 \cdot 19 \\ &= \mathbf{1 \cdot 2^2 \cdot 3^3 \cdot 5^2 \cdot 7^2 \cdot 11 \cdot 13 \cdot 19}\end{aligned}$$

Simplifying fractions

A fraction represents the division between a numerator N (the dividend) and a denominator D (the divisor). If, after decomposing N and D into factors, some of them are repeated in both, we can eliminate them directly and thus reduce the fraction, since, in general:

$$\frac{N}{D} = \frac{A \cdot \boldsymbol{c}}{B \cdot \boldsymbol{c}} = \frac{A}{B} \cdot \frac{\boldsymbol{c}}{\boldsymbol{c}} = \frac{A}{B} \cdot \mathbf{1} = \frac{A}{B}$$

where $A \cdot \boldsymbol{c}$ es N in factors, $B \cdot \boldsymbol{c}$ is D in factors, $\boldsymbol{c}$ is the number shared by both, and A/B is the resulting **equivalent fraction**, which means that:

$$\frac{\boldsymbol{N}}{\boldsymbol{D}} = \frac{\boldsymbol{A}}{\boldsymbol{B}} \Leftrightarrow \frac{N}{D} - \frac{A}{B} = \frac{N \cdot B - D \cdot A}{D \cdot B} = 0 \Leftrightarrow \boldsymbol{N} \cdot \boldsymbol{B} = \boldsymbol{D} \cdot \boldsymbol{A}$$

For example:

$$\frac{N}{D} = \frac{145}{2465} = \frac{29 \cdot \mathbf{5}}{493 \cdot \mathbf{5}} = \frac{\mathbf{29}}{17 \cdot \mathbf{29}} = \frac{1}{17}$$

and

$$\frac{145}{2465} = \frac{1}{17} \Leftrightarrow \mathbf{145} \cdot \mathbf{17} = 2465 = \mathbf{2465} \cdot \mathbf{1}$$

Scandalous Simplification of Fractions

If we have two equivalent fractions, each one with the same number of figures in the numerator and denominator, although not necessarily identical in both, they are equivalent to another whose numerator and denominator are respectively the ordered fusion of the numerators and denominators of the fractions of origin as many times as we want, wherever we want, that is:

$$\frac{A}{B}=\frac{C}{D}\Leftrightarrow\frac{A}{B}=\frac{C}{D}=\frac{A\boldsymbol{C}}{B\boldsymbol{D}}=\frac{C\boldsymbol{A}}{D\boldsymbol{B}}=\frac{A\boldsymbol{A}}{B\boldsymbol{B}}=\frac{CC}{DD}=\frac{A\boldsymbol{C}C}{B\boldsymbol{D}D}=\cdots$$

where A and B have the same number of figures, as well as C and D (it's necessary to add zeros to the left to adjust), although each fraction can have a different one:

$$\frac{11}{3}=\frac{220}{60}\Leftrightarrow\frac{11}{3}=\frac{220}{60}=\frac{11\mathbf{220}}{3\mathbf{060}}=\frac{220\mathbf{11}}{60\mathbf{03}}=\frac{11\mathbf{11}}{3\mathbf{03}}=\cdots$$

This beautiful and powerful theorem devised in my youth, is valid for any number of fractions, e.g.:

$$\frac{1}{2}=\frac{7}{14}=\frac{9}{18}\Leftrightarrow\frac{1}{2}=\frac{1\mathbf{07}09}{2\mathbf{14}18}=\frac{9\mathbf{1}07}{18\mathbf{2}14}=\frac{7\mathbf{09}}{14\mathbf{18}}=\cdots$$

Why it works is amazingly simple:

The necessary and sufficient condition for the A/B and C/D fractions to be equivalent is:

$$\frac{A}{B} = \frac{C}{D} \Leftrightarrow \boldsymbol{A \cdot D = B \cdot C \Leftrightarrow D \cdot A = C \cdot B}$$

Considering this, these simple calculations are enough to deduce the equivalence between CA/DB and A/B:

$$\frac{CA}{DB} \stackrel{\text{def}}{=} \frac{10 \cdot C + A}{10 \cdot D + B} = \frac{A}{B} \Leftrightarrow (10 \cdot \boldsymbol{C} + \boldsymbol{A}) \cdot \boldsymbol{B} = (10 \cdot \boldsymbol{D} + \boldsymbol{B}) \cdot \boldsymbol{A}$$

and we conclude by seeing that the equality is true:

$$\begin{aligned}(10 \cdot \boldsymbol{C} + \boldsymbol{A}) \cdot \boldsymbol{B} &= 10 \cdot \boldsymbol{C} \cdot \boldsymbol{B} + A \cdot B = 10 \cdot \boldsymbol{D} \cdot \boldsymbol{A} + A \cdot B \\ &= 10 \cdot D \cdot A + B \cdot A = (10 \cdot \boldsymbol{D} + \boldsymbol{B}) \cdot \boldsymbol{A}\end{aligned}$$

The equivalence between CA/DB and C/D is similar:

$$\frac{CA}{DB} \stackrel{\text{def}}{=} \frac{10 \cdot C + A}{10 \cdot D + B} = \frac{C}{D} \Leftrightarrow (10 \cdot \boldsymbol{C} + \boldsymbol{A}) \cdot \boldsymbol{D} = (10 \cdot \boldsymbol{D} + \boldsymbol{B}) \cdot \boldsymbol{C}$$

and we conclude by seeing that the equality is true:

$$\begin{aligned}(10 \cdot \boldsymbol{C} + \boldsymbol{A}) \cdot \boldsymbol{D} &= 10 \cdot C \cdot D + \boldsymbol{A} \cdot \boldsymbol{D} = 10 \cdot C \cdot D + \boldsymbol{B} \cdot \boldsymbol{C} \\ &= 10 \cdot D \cdot C + B \cdot C = (10 \cdot \boldsymbol{D} + \boldsymbol{B}) \cdot \boldsymbol{C}\end{aligned}$$

Formulas embedded in fractions

If we know or intuit what the sum of a series of numbers is, we can use the scandalous simplification of the previous point by forming a fraction that merges the result of the sum with its forecast. E.g., the first nonnegative integers can be added from outside to inside taken two at a time (starting at 0 if $n\ is\ odd$ and at 1 if $n\ is\ even$). If $n\ is\ odd$:

$$\mathbf{0+1+2+3+\cdots+n} = (0+n) + \big(1+(n-1)\big) + \cdots$$
$$= (n)\cdot \underbrace{\cdots}_{(n+1)/2} \cdot (n) = \mathbf{\frac{n(n+1)}{2}}$$

And if $n\ is\ even$:

$$\mathbf{1+2+3+\cdots+n} = (1+n) + \big(2+(n-1)\big) + \cdots$$
$$= (n+1)\cdot \underbrace{\cdots}_{n/2} \cdot (n+1) = \mathbf{\frac{n(n+1)}{2}}$$

In both cases:

$$\mathbf{1+2+3+\cdots+n = \frac{n(n+1)}{2}}$$

The following fraction *(true if and only if the previous calculations are verified)* contains the previous formula:

$$\frac{[1+2+3+\cdots+n][n(n+1)]}{10\underbrace{\ldots}_{m}02} = \frac{1+2+3+\cdots+n}{1}$$

$$= \mathbf{1+2+3+\cdots+n} = \frac{\boldsymbol{n(n+1)}}{\mathbf{2}} = \frac{n(n+1)}{0\underbrace{\ldots}_{m}02}$$

where m is the number of digits of $n(n+1)$ minus one and the brackets mean *to operate and put the result.*

Example 1. The sum of the numbers from 1 to 5 is 15 (half the product of 5 *times* 6); the resulting fraction is:

$$\frac{[1+2+3+4+5][5(5+1)]}{10\underbrace{\ldots}_{2-1}02} = \frac{15\mathbf{30}}{\mathbf{102}} = \frac{15}{1} = 15 = \frac{30}{2}$$

Example 2. The sum of the numbers from 1 to 22 is 253 (half of 506 — the product of 22 *times* 23); the resulting fraction is:

$$\frac{[1+\cdots+22][22(22+1)]}{10\underbrace{\ldots}_{3-1}02} = \frac{253\mathbf{506}}{\mathbf{1002}} = \frac{253}{1} = 253 = \frac{506}{2}$$

Significant figures and rounding

A significant figure is any digit in the number that is not a leading or trailing zero. The intermediate zeros are significant, since they modify the weighting; 1.02 is not the same as 1.2:

$$1.02 = \mathbf{1} \cdot 10^0 + \mathbf{0} \cdot 10^{-1} + \mathbf{2} \cdot 10^{-2}$$

and

$$1.2 = 1.20 = \mathbf{1} \cdot 10^0 + \mathbf{2} \cdot 10^{-1} + \mathbf{0} \cdot 10^{-2}$$

The zeros used to locate the decimal point position are not significant figures; neither are zeros at the beginning of an integer.

For **example**, 0.0123 contains only three significant figures, since the two zeros serve to place the decimal point, which is best seen in scientific notation ($1.23 \cdot 10^{-2}$). The number $7.0 \cdot 10^2$ has two significant figures and $7.00 \cdot 10^2$ has three, even though 700 represents both. The same number with 4 significant figures is 700.0 (in scientific notation $7.000 \cdot 10^2$). Both 0.34 and .34 have two significant figures; the leading 0 is ignored. The same happens with the numbers 034 and 34.

The **conventions for rounding** are as follows:

- When the number to remove is *less than* 5, the one before it does not change (e.g., 5.2**4** is rounded to 5.2).
- When the number is *greater than* 5, the one before it is increased by 1 (e.g., 5.2**6** rounds to 5.3).
- When it is 5, the number that precedes it is not changed if it is even (0 is considered even), but if it is odd, it is increased by 1 (e.g., 5.2**5** is rounded to 5.2 and 2.3**5** rounds to 2.4).

Rounding in operations:

- In *multiplication and division*, the result must not have more significant figures than the smallest number of significant figures used in the operation. E.g., «$1.25 \cdot 2.2$» results in 2.75, which rounded to two significant figures is 2.8.
- In *addition and subtraction,* the position of the first suspect digit determines the last digit that is retained. E.g., the sum «$7.2\mathbf{4} + 2.\mathbf{3}$» is 9.54 and is expressed as 9.5; The subtraction «$7.2\mathbf{7} - 2.234\mathbf{3}$» is 5.0357 and is expressed as 5.04.

Chapter 10

Calculation of logarithms

The logarithm is a mathematical tool that allows us to obtain the number n to which b must be raised (the *base* of the logarithm, positive and different from 1) to obtain a (a positive real number —*the antilogarithm or argument*—):

$$\log_b a = \boldsymbol{n} \Leftrightarrow b^n = a$$

The common logarithm (***log***) uses the number 10 and the natural or Napierian logarithm (***ln***) the number e:

$$\log_{10} a = \mathbf{log}\, a \quad and \quad \log_e a = \mathbf{ln}\, a$$

It's possible to use other bases; in particular in this chapter, we will extensively use binary (*base* 2) as it facilitates manual calculations.

Calculation of the number *e*

This number is immeasurable (it cannot be expressed as a fraction, but it can be expressed as an infinite sum of them); its approximate value is 2.718281828459...

The number e is defined as the limit to which the expression $\left(1+\frac{1}{n}\right)^n$ tends when n becomes infinite. We know that Newton's binomial looks like this:

$$(x+a)^m = \binom{m}{0} x^m + \cdots + \binom{m}{n} x^{m-n} a^n + \cdots + \binom{m}{m} a^m$$

where

$$\binom{\boldsymbol{m}}{\boldsymbol{n}} = \frac{\boldsymbol{m(m-1)\ldots(m-n+1)}}{\boldsymbol{1\cdot 2\cdot 3\cdot \cdots \cdot n}} = \frac{m(m-1)\ldots(m-n+1)}{n!}$$
$$= \frac{m\ldots\big(m-(n-1)\big)(m-n)!}{n!\,(m-n)!} = \frac{\boldsymbol{m!}}{\boldsymbol{n!\,(m-n)!}}$$
$$= \frac{m!}{(m-n)!\,n!} = \frac{m!}{(m-n)!\,\big(m-(m-n)\big)!}$$
$$= \binom{\boldsymbol{m}}{\boldsymbol{m-n}}$$

and in particular,

$$\binom{\boldsymbol{m}}{\boldsymbol{0}} = \frac{m!}{(m-0)!\,\big(m-(m-0)\big)!} = \frac{m!}{m!\,(m-m)!} = \binom{\boldsymbol{m}}{\boldsymbol{m}} = \boldsymbol{1}$$

Developing the binomial of the number e:

$$\left(1+\frac{1}{n}\right)^n = \frac{1}{0!}\cdot 1^n + \frac{n}{1!}\cdot 1^{n-1}\cdot\left(\frac{1}{n}\right)^1 + \frac{n(n-1)}{2!}\cdot 1^{n-2}\cdot\left(\frac{1}{n}\right)^2 + \cdots + \frac{n(n-1)\ldots(n-(n-1))}{n!}\cdot 1^{n-n}\cdot\left(\frac{1}{n}\right)^n$$

simplifying and rearranging denominators:

$$\left(1+\frac{1}{n}\right)^n = \frac{1}{0!}+\frac{1}{1!}+\frac{n(n-1)}{n^2}\cdot\frac{1}{2!}+\frac{n(n-1)(n-2)}{n^3}\cdot\frac{1}{3!}+\cdots + \frac{n(n-1)\ldots(n-n+1)}{n^n}\cdot\frac{1}{n!}$$

$$= \frac{1}{0!}+\frac{1}{1!}+\frac{n-1}{n}\cdot\frac{1}{2!}+\frac{n-1}{n}\cdot\frac{n-2}{n}\cdot\frac{1}{3!}+\cdots + \frac{n-1}{n}\cdot\frac{n-2}{n}\cdot\ldots\cdot\frac{n-n+1}{n}\cdot\frac{1}{n!}$$

and performing operations:

$$\left(1+\frac{1}{n}\right)^n = \frac{1}{0!}+\frac{1}{1!}+\left(1-\frac{1}{n}\right)\cdot\frac{1}{2!}+\left(1-\frac{1}{n}\right)\cdot\left(1-\frac{2}{n}\right)\cdot\frac{1}{3!}+\cdots + \left(1-\frac{1}{n}\right)\cdot\left(1-\frac{2}{n}\right)\cdot\ldots\cdot\left(1-\frac{n-1}{n}\right)\cdot\frac{1}{n!}$$

When n is so large that it tends to be infinite, the fractions that have it as a denominator tend to be zero, that is:

$$\left(\mathbf{1}+\frac{\mathbf{1}}{\boldsymbol{n}}\right)^{n}=\frac{1}{0!}+\frac{1}{1!}+(1-0)\cdot\frac{1}{2!}+(1-0)\cdot(1-0)\cdot\frac{1}{3!}+\cdots$$
$$+(1-0)\cdot(1-0)\cdot\cdots\cdot(1-0)\cdot\frac{1}{n!}+\cdots$$
$$=\mathbf{1}+\mathbf{1}+\frac{\mathbf{1}}{\mathbf{2!}}+\frac{\mathbf{1}}{\mathbf{3!}}+\cdots+\frac{\mathbf{1}}{\boldsymbol{n}!}+\cdots$$

The sum to infinity of all the terms is the number e; the more addends used, the better the approximation. Let's add the first terms up to $n = 16$:

Partial sum	n	$1/_{n!}$
1	0	1
2	1	1
2.5	2	0.5
2.66666666666667	3	0.16666666666667
2.70833333333334	4	0.04166666666667
2.71666666666667	5	0.00833333333333
2.71805555555556	6	0.00138888888889
2.71825396825397	7	0.00019841269841
2.71827876984127	8	0.00002480158730
2.71828152557319	9	0.00000275573192
2.71828180114638	10	0.00000027557319
2.71828182619849	11	0.00000002505211
2.71828182828617	12	0.00000000208768
2.71828182844676	13	0.00000000016059
2.71828182845823	14	0.00000000001147
2.71828182845899	15	0.00000000000076
2.71828182845904	16	0.00000000000005

(*approximate value*) $e \cong$ 2.71828182845904
(16 *exact decimals*) $e =$ 2.7182818284590452...

Calculating the base 10 logarithm of 2

The logarithm of a number consists of an integer part *(the characteristic)* and a fractional part *(the mantissa)*. The exact value of the decimal ($base$ 10) or common logarithm of 2 can be rounded off with virtually no loss of precision:

$$\log_{10} 2 = \log 2 = 0.30102999566398 \cong 0.30103$$

To find the **characteristic** of the common logarithm of a number a we look to see if a is less than 10; if it is, the characteristic is 0; if not, it is $divided\ by$ 10 over and over again until the result of one of the operations is, always increasing the characteristic by 1 on each attempt, for example, assuming that $\log a$ has characteristic 3 the sequence of operations would be:

$Calculation$	$Verification$	$\boldsymbol{Characteristic}$
a	$a > 10$	(0)
$a/10 = c_1$	$c_1 > 10$	(1)
$c_1/10 = c_2$	$c_2 > 10$	(2)
$c_2/10 = \boldsymbol{c_3}$	$\boldsymbol{c_3} < 10$	$(\log a = \mathbf{3}. \ldots)$

In this case, the **logarithm of 2** has **characteristic $\mathbf{0}$** since 2 is less than 10, that is, $\log 2 = \mathbf{0}. \ldots$

To find the **mantissa** of the common logarithm of a number a, we take the last result of the calculation of the characteristic that determined it (c_3) and after raising it to the tenth power ($m_{00} = {c_3}^{10}$) we check if it is less than ten ($m_{00} < 10$); if it is, the first digit of the mantissa is 0; otherwise, we must divide it by 10 over and over again $\left(m_{0j} = m_{0(j-1)}/10\right)$ until the result of one of the operations is ($m_{0k} < 10$) in which case the first digit of the mantissa (m_0) is equal to the iteration counter ($m_0 = k$). To find the next digits of the mantissa, we have to repeat the same procedure starting from the last result that determined the previous digit of the mantissa raised to the tenth power, namely $m_{i0} = {m_{(i-1)k}}^{10}$, where i is the current position of the mantissa digit, and $m_{(i-1)k} < 10$. The algorithm is the same; we check if $m_{i0} < 10$; if it is, the i digit of the mantissa is 0; otherwise, we divide by 10 over and over again $\left(m_{ij} = m_{i(j-1)}/10\right)$ until the result of one of the operations is ($m_{ik'} < 10$) in which case the digit i of the mantissa (m_i) is the current iteration number ($m_i = k'$).

Raising a number to the tenth power consists of *transforming its ninth power (its cube to the cube) into the next one* as we saw in previous chapters. In this case, we first need to cube 2 and *cube the result again*, de nuevo *al cubo*, which doesn't require much computation:

$$2^3 = 2 \cdot 2 \cdot 2 = 8$$
$$8^3 = 8 \cdot 8^2 = 8 \cdot 64 = 8 \cdot (60 + 4) = \mathbf{480} + \mathbf{32} = \mathbf{51}2$$
$$2^9 = (2^3)^3 = 8^3 = 512$$

and then we add the necessary addends that transform that power into the following (or we just calculate 512 *times* 2):

$$\begin{array}{rcccc} \mathbf{2}^9 = & & 5 & 1 & 2 \\ (\mathbf{2}-1)\cdot 5 = & 0 & 5 & & \\ (\mathbf{2}-1)\cdot 1 = & & 0 & 1 & \\ (\mathbf{2}-1)\cdot 2 = & & & 0 & 2 \\ + & - & -- & - & - \\ \mathbf{2}^{10} = & \mathbf{1} & \mathbf{0} & \mathbf{2} & \mathbf{4} \end{array}$$

We can find **the first digit of the mantissa** like this:

$$\begin{array}{rcc} \textit{Calculation} & \textit{Verification} & \boldsymbol{Mantissa} \\ 1024 & 1024 > 10 & (0) \\ 1024/10 = 102.4 & 102.4 > 10 & (1) \\ 102.4/10 = 10.24 & 10.24 > 10 & (2) \\ 10.24/10 = \mathbf{1.024} & \mathbf{1.024} < 10 & (\log 2 = 0.\mathbf{3}...) \end{array}$$

The second digit of the mantissa requires *raising* the last result (1.024) *to* 10; to do this, we first calculate its square and from it its cube:

$$
\begin{array}{r c c c c c c c c c}
\mathbf{1.0}^{\mathbf{2}} = & 1. & 0 & 0 & & & & & & \\
2\cdot(\mathbf{2}\cdot 1) = & & 0 & 4 & & & & & & \\
2\cdot(\mathbf{2}\cdot 0) = & & & 0 & 0 & & & & & \\
\mathbf{2}^{2} = & & & & 0 & 4 & & & & \\
+ & - & - & - & - & - & & & & \\
\mathbf{1.02}^{\mathbf{2}} = & \mathbf{1.} & \mathbf{0} & \mathbf{4} & \mathbf{0} & \mathbf{4} & & & & \\
2\cdot(\mathbf{4}\cdot\mathbf{1}) = & & & \mathbf{0} & \mathbf{8} & & & & & \\
2\cdot(\mathbf{4}\cdot\mathbf{0}) = & & & & \mathbf{0} & \mathbf{0} & & & & \\
2\cdot(\mathbf{4}\cdot\mathbf{2}) = & & & & & \mathbf{1} & \mathbf{6} & & & \\
\mathbf{4}^{\mathbf{2}} = & & & & & & \mathbf{1} & \mathbf{6} & & \\
+ & g & f & e & d & c & b & a & & \\
\mathbf{1.024}^{\mathbf{2}} = & 1, & 0 & 4 & 8 & 5 & 7 & 6 & & \\
(\mathbf{10}-1)g = & 09 & & & & & & & & \\
(\mathbf{10}-1)f+\mathbf{2}g = & 0 & 2 & & & & & & & \\
(\mathbf{10}-1)e+\mathbf{2}f+\mathbf{4}g = & & 4 & 0 & & & & & & \\
(\mathbf{10}-1)d+\mathbf{2}e+\mathbf{4}f = & & & 8 & 0 & & & & & \\
(\mathbf{10}-1)c+\mathbf{2}d+\mathbf{4}e = & & & & 7 & 7 & & & & \\
(\mathbf{10}-1)b+\mathbf{2}c+\mathbf{4}d = & & & & 1 & 0 & 5 & & & \\
(\mathbf{10}-1)a+\mathbf{2}b+\mathbf{4}c = & & & & & & 8 & 8 & & \\
\mathbf{2}a+\mathbf{4}b = & & & & & & & 4 & 0 & \\
\mathbf{4}a = & & & & & & & & 2 & 4 \\
+ & - & - & - & - & - & - & - & - & - \\
\mathbf{1.024}^{\mathbf{3}} = & \mathbf{1.0} & \mathbf{7} & \mathbf{3} & \mathbf{7} & \mathbf{4} & \mathbf{1} & \mathbf{8} & \mathbf{2} & \mathbf{4}
\end{array}
$$

Now we have to find the ninth power of 1.024 by obtaining the cube of the previous result from its square, by adding the corresponding terms. To obtain the square of 1.073 741 824, we calculate the squares of 10, 107, 1073, 10737, 107374, 1073741, 10737418, 107374182 and 1073741824, transforming the previous one into the next one by adding a few terms (the number of figures in the square to be transformed determines the position of the first of them and the others are moved one place to the right); in the operations the decimal point is not taken into account until the final result:

$10^2 =$	1	0	0								
$2 \cdot (7 \cdot 1) =$		1	4								
$2 \cdot (7 \cdot 0) =$			0	0							
$7^2 =$				4	9						
$107^2 +=$	**1**	**1**	**4**	**4**	**9**						
$2 \cdot (3 \cdot 1) =$			**0**	**6**							
$2 \cdot (3 \cdot 0) =$				**0**	**0**						
$2 \cdot (3 \cdot 7) =$					**4**	**2**					
$3^2 =$						**0**	**9**				
$1073^2 +=$	1	1	5	1	3	2	9				
$2 \cdot (7 \cdot 1) =$				1	4						
$2 \cdot (7 \cdot 0) =$					0	0					
$2 \cdot (7 \cdot 7) =$						9	8				
$2 \cdot (7 \cdot 3) =$							4	2			
$7^2 =$								4	9		
$10737^2 +=$	**1**	**1**	**5**	**2**	**8**	**3**	**1**	**6**	**9**		
$2 \cdot (4 \cdot 1) =$					**0**	**8**					
$2 \cdot (4 \cdot 0) =$						**0**	**0**				
$2 \cdot (4 \cdot 7) =$							**5**	**6**			
$2 \cdot (4 \cdot 3) =$								**2**	**4**		
$2 \cdot (4 \cdot 7) =$									**5**	**6**	
$4^2 =$										**1**	**6**
$1.073\,74^2 +=$	1.	1	5	2	9	1	7	5	8	7	6

As the left part of the number is becoming fixed, we are going to continue with the squares that remain to be calculated in another table that links to the previous one; the ellipsis points represent the number 1.1529, that is, the first part of the square of 1.073 74:

$$
\begin{array}{rclcccccccccc}
\mathbf{1.073\,74^2} & = & \ldots & 1 & 7 & 5 & 8 & 7 & 6 & & & & \\
2\cdot(\mathbf{1}\cdot\mathbf{1}) & = & & 0 & 2 & & & & & & & & \\
2\cdot(\mathbf{1}\cdot\mathbf{0}) & = & & & 0 & 0 & & & & & & & \\
2\cdot(\mathbf{1}\cdot\mathbf{7}) & = & & & & 1 & 4 & & & & & & \\
2\cdot(\mathbf{1}\cdot\mathbf{3}) & = & & & & & 0 & 6 & & & & & \\
2\cdot(\mathbf{1}\cdot\mathbf{7}) & = & & & & & & 1 & 4 & & & & \\
2\cdot(\mathbf{1}\cdot\mathbf{4}) & = & & & & & & & 0 & 8 & & & \\
\mathbf{1^2} & = & & & & & & & & 0 & 1 & & \\
 & + & - & - & - & - & - & - & - & - & - & & \\
\mathbf{1.073\,741^2} & = & \ldots & \mathbf{1} & \mathbf{9} & \mathbf{7} & \mathbf{3} & \mathbf{5} & \mathbf{0} & \mathbf{8} & \mathbf{1} & & \\
2\cdot(\mathbf{8}\cdot\mathbf{1}) & = & & & \mathbf{1} & \mathbf{6} & & & & & & & \\
2\cdot(\mathbf{8}\cdot\mathbf{0}) & = & & & & \mathbf{0} & \mathbf{0} & & & & & & \\
2\cdot(\mathbf{8}\cdot\mathbf{7}) & = & & & & \mathbf{1} & \mathbf{1} & \mathbf{2} & & & & & \\
2\cdot(\mathbf{8}\cdot\mathbf{3}) & = & & & & & & \mathbf{4} & \mathbf{8} & & & & \\
2\cdot(\mathbf{8}\cdot\mathbf{7}) & = & & & & & & \mathbf{1} & \mathbf{1} & \mathbf{2} & & & \\
2\cdot(\mathbf{8}\cdot\mathbf{4}) & = & & & & & & & & \mathbf{6} & \mathbf{4} & & \\
2\cdot(\mathbf{8}\cdot\mathbf{1}) & = & & & & & & & & & \mathbf{1} & \mathbf{6} & \\
\mathbf{8^2} & = & & & & & & & & & & \mathbf{6} & \mathbf{4} \\
 & + & - & - & - & - & - & - & - & - & - & - & - \\
\mathbf{1.073\,741\,8^2} & = & \ldots & 2 & 1 & 4 & 5 & 3 & 0 & 6 & 7 & 2 & 4
\end{array}
$$

In the following table the ellipses represent 1.152 92 (the beginning of the square of 1.073 741 8):

$\mathbf{1.073\,741\,8^2} =$	...	1	4	5	3	0	6	7	2	4		
$2 \cdot (\mathbf{2} \cdot \mathbf{1}) =$			0	4								
$2 \cdot (\mathbf{2} \cdot \mathbf{0}) =$				0	0							
$2 \cdot (\mathbf{2} \cdot \mathbf{7}) =$					2	8						
$2 \cdot (\mathbf{2} \cdot \mathbf{3}) =$						1	2					
$2 \cdot (\mathbf{2} \cdot \mathbf{7}) =$							2	8				
$2 \cdot (\mathbf{2} \cdot \mathbf{4}) =$								1	6			
$2 \cdot (\mathbf{2} \cdot \mathbf{1}) =$									0	4		
$2 \cdot (\mathbf{2} \cdot \mathbf{8}) =$										3	2	
$\mathbf{2^2} =$											0	4
+	–	–	–	–	–	–	–	–	–	–	–	–
$\mathbf{1.073\,741\,82^2} =$	...	**1**	**4**	**9**	**6**	**0**	**1**	**6**	**9**	**1**	**2**	**4**

In the following table, the ellipses represent 1.152 921 (the beginning of the square of 1.073 741 82):

$\mathbf{1.073\,741\,82^2} =$	...	**4**	**9**	**6**	**0**	**1**	**6**	**9**	**1**	**2**	**4**		
$2 \cdot (\mathbf{4} \cdot \mathbf{1}) =$			**0**	**8**									
$2 \cdot (\mathbf{4} \cdot \mathbf{0}) =$				**0**	**0**								
$2 \cdot (\mathbf{4} \cdot \mathbf{7}) =$					**5**	**6**							
$2 \cdot (\mathbf{4} \cdot \mathbf{3}) =$						**2**	**4**						
$2 \cdot (\mathbf{4} \cdot \mathbf{7}) =$							**5**	**6**					
$2 \cdot (\mathbf{4} \cdot \mathbf{4}) =$								**3**	**2**				
$2 \cdot (\mathbf{4} \cdot \mathbf{1}) =$									**0**	**8**			
$2 \cdot (\mathbf{4} \cdot \mathbf{8}) =$										**6**	**4**		
$2 \cdot (\mathbf{4} \cdot \mathbf{2}) =$											**1**	**6**	
$\mathbf{4^2} =$												**1**	**6**
+	–	–	–	–	–	–	–	–	–	–	–	–	–
$\mathbf{1.073\,741\,824^2} =$	...	5	0	4	6	0	6	8	4	6	9	7	6

that is, $1.024^6 = (1.024^3)^2 = 1.152\,921\,504\,606\,846\,976$, which can be rounded to the ninth decimal place without losing much precision, being simplified to 1.152 921 505 and thus reducing the calculations of conversion to cube, with which we would have $(1{,}024^3)^3 = 1{,}024^9$. The addends necessary for the conversion from one power to the next are (considering the last decimal place without rounding):

$$\mathbf{1\,073\,741\,824^2} = \begin{matrix} 1 & 1 & 5 & 2 & 9 & 2 & 1 & 5 & 0 & 4 \\ j & i & h & g & f & e & d & c & b & a \end{matrix}$$

$$\boxed{A} = (\mathbf{1}-1)j = 000$$
$$\boxed{B} = (\mathbf{1}-1)i + \mathbf{0}j = 000$$
$$\boxed{C} = (\mathbf{1}-1)h + \mathbf{0}i + \mathbf{7}j = 007$$
$$\boxed{D} = (\mathbf{1}-1)g + \mathbf{0}h + \mathbf{7}i + \mathbf{3}j = 010$$
$$\boxed{E} = (\mathbf{1}-1)f + \mathbf{0}g + \mathbf{7}h + \mathbf{3}i + \mathbf{7}j = 045$$
$$\boxed{F} = (\mathbf{1}-1)e + \mathbf{0}f + \mathbf{7}g + \mathbf{3}h + \mathbf{7}i + \mathbf{4}j = 040$$
$$\boxed{G} = (\mathbf{1}-1)d + \mathbf{0}e + \mathbf{7}f + \mathbf{3}g + \mathbf{7}h + \mathbf{4}i + \mathbf{1}j = 109$$
$$\boxed{H} = (\mathbf{1}-1)c + \mathbf{0}d + \mathbf{7}e + \mathbf{3}f + \mathbf{7}g + \mathbf{4}h + \mathbf{1}i + \mathbf{8}j = 084$$
$$\boxed{I} = (\mathbf{1}-1)b + \mathbf{0}c + \mathbf{7}d + \mathbf{3}e + \mathbf{7}f + \mathbf{4}g + \mathbf{1}h + \mathbf{8}i + \mathbf{2}j = 099$$
$$(\mathbf{1}-1)a + \mathbf{0}b + \mathbf{7}c + \mathbf{3}d + \mathbf{7}e + \mathbf{4}f + \mathbf{1}g + \mathbf{8}h + \mathbf{2}i + \mathbf{4}j = 136$$
$$\boxed{K} = \mathbf{0}a + \mathbf{7}b + \mathbf{3}c + \mathbf{7}d + \mathbf{4}e + \mathbf{1}f + \mathbf{8}g + \mathbf{2}h + \mathbf{4}i = 069$$
$$\boxed{L} = \mathbf{7}a + \mathbf{3}b + \mathbf{7}c + \mathbf{4}d + \mathbf{1}e + \mathbf{8}f + \mathbf{2}g + \mathbf{4}h = 165$$
$$\boxed{M} = \mathbf{3}a + \mathbf{7}b + \mathbf{4}c + \mathbf{1}d + \mathbf{8}e + \mathbf{2}f + \mathbf{4}g = 075$$
$$\boxed{N} = \mathbf{7}a + \mathbf{4}b + \mathbf{1}c + \mathbf{8}d + \mathbf{2}e + \mathbf{4}f = 081$$
$$\boxed{O} = \mathbf{4}a + \mathbf{1}b + \mathbf{8}c + \mathbf{2}d + \mathbf{4}e = 066$$
$$\boxed{P} = \mathbf{1}a + \mathbf{8}b + \mathbf{2}c + \mathbf{4}d = 018$$
$$\boxed{Q} = \mathbf{8}a + \mathbf{2}b + \mathbf{4}c = 052$$
$$\boxed{R} = \mathbf{2}a + \mathbf{4}b = 008$$
$$\boxed{S} = \mathbf{4}a = 016$$

which, properly located and added from top to bottom

and from left to right, transform the square of 1.073 741 824 into its cube, as well as the square of 1.024^3 into its cube and therefore the sixth power of 1.024 in its power to the ninth (the square has been rounded to the eleventh decimal place; we only needed the first few calculations in the table above):

$\mathbf{1.073\,741\,824^2}$	=	1.	1	5	2	9	2	1	5	0	4	6	1	
$\boxed{A}$	=	0												
$\boxed{B}$	=	0	0											
$\boxed{C}$	=	0	0	7										
$\boxed{D}$	=		0	1	0									
$\boxed{E}$	=			0	4	5								
$\boxed{F}$	=				0	4	0							
$\boxed{G}$	=					1	0	9						
$\boxed{H}$	=						0	8	4					
$\boxed{I}$	=							0	9	9				
$\boxed{J}$	=								1	3	6			
$\boxed{K}$	=									0	6	9		
$\boxed{L}$	=										1	6	5	
$\boxed{M}$	=											0	7	5
$\mathbf{1.073\,741\,824^3}$	=	**1.**	**2**	**3**	**7**	**9**	**4**	**0**	**0**	**3**	**9**	**2**	**3**	...
$(1.024^3)^3$	=	1.	2	3	7	9	4	0	0	3	9	2	8	5

The last line of the previous table shows the real value of 1.024^9, which compared to the cube conversion that we have obtained, decides a value of 1.237 940 039 3 for the ninth power of 1024.

Now we have to obtain 1024^{10} by multiplying the previous result by 1024:

$$1.024^{10} = 1.024 \cdot 1.237\,940\,039\,3 = 1.267\,650\,600\,2$$

or transforming 1.024^9 into its next power by adding the following conversion terms

$$\mathbf{1024^9} = \begin{matrix} 1 & 2 & 3 & 7 & 9 & 4 & 0 & 0 & 3 & 9 \\ j & i & h & g & f & e & d & c & b & a \end{matrix}$$

$$\boxed{A} = (\mathbf{1} - 1)j = 00$$

$$\boxed{B} = (\mathbf{1} - 1)i + \mathbf{0}j = 00$$

$$\boxed{C} = (\mathbf{1} - 1)h + \mathbf{0}i + \mathbf{2}j = 02$$

$$\boxed{D} = (\mathbf{1} - 1)g + \mathbf{0}h + \mathbf{2}i + \mathbf{4}j = 08$$

$$\boxed{E} = (\mathbf{1} - 1)f + \mathbf{0}g + \mathbf{2}h + \mathbf{4}i = 14$$

$$\boxed{F} = (\mathbf{1} - 1)e + \mathbf{0}f + \mathbf{2}g + \mathbf{4}h = 26$$

$$\boxed{G} = (\mathbf{1} - 1)d + \mathbf{0}e + \mathbf{2}f + \mathbf{4}g = 46$$

$$\boxed{H} = (\mathbf{1} - 1)c + \mathbf{0}d + \mathbf{2}e + \mathbf{4}f = 44$$

$$\boxed{I} = (\mathbf{1} - 1)b + \mathbf{0}c + \mathbf{2}d + \mathbf{4}e = 16$$

$$\boxed{J} = (\mathbf{1} - 1)a + \mathbf{0}b + \mathbf{2}c + \mathbf{4}d = 00$$

$$\boxed{K} = \mathbf{0}a + \mathbf{2}b + \mathbf{4}c = 06$$

$$\boxed{L} = \mathbf{2}a + \mathbf{4}b = 30$$

$$\boxed{M} = \mathbf{4}a = 36$$

to the ninth power of 1024 duly shifted from top to bottom as indicated in the following table:

$\mathbf{1.024^9} =$	1,	2	3	7	9	4	0	0	3	9	3	...
$\boxed{A} =$	0											
$\boxed{B} =$	0	0										
$\boxed{C} =$		0	2									
$\boxed{D} =$			0	8								
$\boxed{E} =$				1	4							
$\boxed{F} =$					2	6						
$\boxed{G} =$						4	6					
$\boxed{H} =$							4	4				
$\boxed{I} =$								1	6			
$\boxed{J} =$									0	0		
$\boxed{K} =$										0	6	
$\boxed{L} =$											3	0
$\mathbf{1.024^{10}} =$	**1.**	**2**	**6**	**7**	**6**	**5**	**0**	**6**	**0**	**0**	**2**	...

This determines the **second digit of the mantissa**. The **other digits** are calculated in the same way:

Calculation	*Verification*	***Mantissa***
$1.024^{10} =$	$\mathbf{1.267\,650\,600\,2} < 10$	(0.3**0**...)
$1.267\,650\,600\,2^{10} =$	$10.715\,086\,069\,5 > 10$	(0)
$/10 =$	$\mathbf{1.071\,508\,606\,9} < 10$	(0.30**1**...)
$1.071\,508\,606\,9^{10} =$	$\mathbf{1.995\,063\,111\,5} < 10$	(0.301 **0**...)
$1.995\,063\,111\,5^{10} =$	$999.002\,066\,070\,9 > 10$	(0)
$/10 =$	$99.900\,206\,607\,0 > 10$	(1)
$/10 =$	$\mathbf{9.990\,020\,660\,7} < 10$	(0.301 0**2**...)
$9.990\,020\,660\,7^{10} =$	$9\,900\,653\,558.96 > 10$	(0)
$/10 =$	$990\,065\,355.896 > 10$	(1)
⋮	⋮	⋮
$/10 =$	$\mathbf{9.900\,653\,558\,9} < 10$	(0.301 02**9**...)
$9.900\,653\,558\,9^{10} =$	$9\,049\,792\,897.30 > 10$	*etc.*

from where we deduce that:

$$\mathbf{log\,2} = \log_{10} 2 = 0.301\,029\,9 \cong \mathbf{0.301\,03}$$

In conclusion, the **manual calculation** of the **logarithm in *base* 10** of a number *is simple but laborious.*

Calculation of the base 2 logarithm of *e*

The *base* 2 logarithm of e is the number to which 2 must be raised to obtain e (approx. 2.718 281 828 459...):

$$\log_2 e = n \Leftrightarrow 2^n = e = \mathbf{2.718\,281\,828\,459}\,045\,235\,360...$$
$$(n = \mathbf{1.442\,695\,04}0\,888\,963\,407\,359\,924\,681\,0...)$$

The algorithm to calculate it manually is similar to the one we used in the previous point changing 10 by 2, that is, we must divide by 2 instead of by 10 and when the number is less than 2 we have found one of the digits (the number of divisions necessary).

For the calculation of **the characteristic** of the logarithm we must check if $e < 2$; it's not, so we divide it by 2; since the result is < 2, the characteristic is **1**:

Calculation	*Verification*	***Characteristic***
	$2.718\,281\,828\,459 > 2$	(0)
$/2 =$	$\mathbf{1.359\,140\,914\,229\,5} < 2$	$(\log_2 e = \mathbf{1}.\,...)$

The *first digit* of **the mantissa** requires squaring the last result (1.359 140 914 229 5); this can be done as before, ignoring the comma:

$$
\begin{array}{r ccccccccc}
\mathbf{1}^2 = & 1 & & & & & & & & \\
2\cdot(\mathbf{3}\cdot\mathbf{1}) = & 0 & 6 & & & & & & & \\
\mathbf{3}^2 = & & 0 & 9 & & & & & & \\
\mathbf{13}^2 \mathrel{+}= & \mathbf{1} & \mathbf{6} & \mathbf{9} & & & & & & \\
2\cdot(\mathbf{5}\cdot\mathbf{1}) = & & \mathbf{1} & \mathbf{0} & & & & & & \\
2\cdot(\mathbf{5}\cdot\mathbf{3}) = & & & \mathbf{3} & \mathbf{0} & & & & & \\
\mathbf{5}^2 = & & & & \mathbf{2} & \mathbf{5} & & & & \\
\mathbf{135}^2 \mathrel{+}= & 1 & 8 & 2 & 2 & 5 & & & & \\
2\cdot(\mathbf{9}\cdot\mathbf{1}) = & & & 1 & 8 & & & & & \\
2\cdot(\mathbf{9}\cdot\mathbf{3}) = & & & & 5 & 4 & & & & \\
2\cdot(\mathbf{9}\cdot\mathbf{5}) = & & & & & 9 & 0 & & & \\
\mathbf{9}^2 = & & & & & & 8 & 1 & & \\
\mathbf{1359}^2 \mathrel{+}= & \mathbf{1} & \mathbf{8} & \mathbf{4} & \mathbf{6} & \mathbf{8} & \mathbf{8} & \mathbf{1} & & \\
2\cdot(\mathbf{1}\cdot\mathbf{1}) = & & & & \mathbf{0} & \mathbf{2} & & & & \\
2\cdot(\mathbf{1}\cdot\mathbf{3}) = & & & & & \mathbf{0} & \mathbf{6} & & & \\
2\cdot(\mathbf{1}\cdot\mathbf{5}) = & & & & & & \mathbf{1} & \mathbf{0} & & \\
2\cdot(\mathbf{1}\cdot\mathbf{9}) = & & & & & & & \mathbf{1} & \mathbf{8} & \\
\mathbf{1}^2 = & & & & & & & & \mathbf{0} & \mathbf{1} \\
\mathbf{13591}^2 \mathrel{+}= & 1 & 8 & 4 & 7 & 1 & 5 & 2 & 8 & 1
\end{array}
$$

and so on until the result is reached:

$$13\,591\,409\,142\,295^2 = 184726402473260107557867025$$

which becomes 1.84726402473260107557867025 when multiplied by $(10^{-13})^2 = 10^{-26}$ to get the place of the decimal point; value that should be rounded to a number of significant figures that preserves the initial precision (we have started using thirteen digits in the representation of the number e) which turns out to be 1.8472640247326. We check if it is less than 2 (if it was not, it would have to be divided by 2 and the digit of the mantissa would be the binary 1); as it is, **the first digit of the mantissa** is $\mathbf{0}$ $(binary)$. To find the other digits of the mantissa we proceed as before; each number that determines each new digit of the mantissa is the one that initiates the next cycle **by squaring it** (operation here denoted $**\,\mathbf{2}$), in this way we obtain a binary sequence of zeros and ones, a result that is transformed to a decimal once the calculation of the logarithm is concluded, multiplying by the corresponding weighting:

$$
\begin{array}{rlll}
 & 1.3591409142295 & Test & Mantissa \\
**\,2 = & 1.8472640247326 & <2 & 0.\mathbf{0}... \\
**\,2 = & 3.4123843770713 & >2 & (0) \\
/2 = & 1.7061921885357 & <2 & 0.0\mathbf{1}... \\
**\,2 = & 2.9110917842202 & >2 & (0) \\
/2 = & 1.4555458921101 & <2 & 0.01\mathbf{1}... \\
**\,2 = & 2.1186138440386 & >2 & (0) \\
/2 = & 1.0593069220193 & <2 & 0.011\mathbf{1}... \\
**\,2 = & 1.1221311550380 & <2 & 0.0111\mathbf{0}... \\
**\,2 = & 1.2591783291069 & <2 & 0.01110\mathbf{0}... \\
**\,2 = & 1.5855300644925 & <2 & 0.011100\mathbf{0}... \\
**\,2 = & 2.5139055854096 & >2 & (0) \\
/2 = & 1.2569527927048 & <2 & 0.0111000\mathbf{1}...
\end{array}
$$

		Test	*Mantissa*
:	1.2569527927048		
** 2 =	1.5799303230884	< **2**	0. ....**0**...
** 2 =	2.4961798258142	> 2	(0)
/2 =	1.2480899129071	< **2**	0. ....0**1**...
** 2 =	1.5577284307005	< **2**	0. ....01**0**...
** 2 =	2.4265178638126	> 2	(0)
/2 =	1.2132589319063	< **2**	0. ....010**1**...
** 2 =	1.4719972358504	< **2**	0. ....0101**0**...
** 2 =	2.1667758623512	> 2	(0)
/2 =	1.0833879311756	< **2**	0. ....01010**1**...
** 2 =	1.1737294094170	< **2**	0. ....010101**0**...
** 2 =	1.3776407265304	< **2**	0. ....0101010**0**...
** 2 =	1.8978939713952	< **2**	0. ....01010100**0**...
** 2 =	3.6020015266582	> 2	(0)
/2 =	1.8010007633291	< **2**	0. ....010101000**1**...
** 2 =	3.2436037495120	> 2	(0)
/2 =	1.6218018747560	< **2**	0. ....0101010001**1**...
** 2 =	2.6302413209621	> 2	(0)
/2 =	1.3151206604811	< **2**	0. ....01010100011**1**...
** 2 =	1.7295423516242	< **2**	0. ....010101000111**0**...
** 2 =	2.9913167460618	> 2	(0)
/2 =	1.4956583730309	< **2**	0. ....0101010001110**1**...
** 2 =	2.2369939688174	> 2	(0)
/2 =	1.1184969844087	< **2**	0. ....01010100011101**1**...
** 2 =	1.2510355041314	< **2**	0. ....010101000111011**0**...
** 2 =	1.5650898325973	> 2	0. ....0101010001110110**0**...
** 2 =	2.4495061840995	> 2	(0)
/2 =	1.2247530920498	< **2**	0. ....01010100011101100**1**...

The conversion of the mantissa to *base* 10 is carried out as follows:

$$0.0\mathbf{1110001010101000111011001}\ldots$$
$$= \mathbf{1} \cdot 2^{-2} + \mathbf{1} \cdot 2^{-3} + \mathbf{1} \cdot 2^{-4} + \mathbf{1} \cdot 2^{-8} + \mathbf{1} \cdot 2^{-10} + \mathbf{1} \cdot 2^{-12} + \mathbf{1} \cdot 2^{-14} + \mathbf{1} \cdot 2^{-18} + \mathbf{1} \cdot 2^{-19} + \mathbf{1} \cdot 2^{-20} + \mathbf{1} \cdot 2^{-22} + \mathbf{1} \cdot 2^{-23} + \mathbf{1} \cdot 2^{-26}$$
$$= \frac{1}{2^2} + \frac{1}{2^3} + \frac{1}{2^4} + \frac{1}{2^8} + \frac{1}{2^{10}} + \frac{1}{2^{12}} + \frac{1}{2^{14}} + \frac{1}{2^{18}} + \frac{1}{2^{19}} + \frac{1}{2^{20}} + \frac{1}{2^{22}} + \frac{1}{2^{23}} + \frac{1}{2^{26}}$$
$$\cong 0.442\,695\,036\,5 \cong \mathbf{0.442\,695\,04}$$

which, together with the characteristic, approximates $\log_2 e$ to $\mathbf{1.442\,695\,04}$; its inverse is $\mathbf{0.693\,147\,181}$.

Calculation of logarithms from $\log_2 a$

The **natural or Napierian logarithm** of a number a can be calculated from the *base* 2 logarithm of a in the following way:

$$\mathbf{ln}\, a = \log_e a = \frac{\log_2 a}{\log_2 e} = (0.693\,147\,181) \cdot \mathbf{log_2}\, a$$

and the **common** or ***base* 10 logarithm** of a number, like this:

$$\mathbf{log}\, a = \log_{10} a = \log_{10} 2 \cdot \log_2 a = (0.30103) \cdot \mathbf{log_2}\, a$$

furthermore, the **natural logarithm** *can be converted to* *the* **common logarithm**:

$$\mathbf{log}\, a = (0.30103) \cdot \mathbf{log_2}\, a = (0.30103) \cdot \frac{\ln a}{0.693147181}$$
$$= (0.434\,294\,5) \cdot \mathbf{ln}\, a$$

and the **common logarithm** in **Napierian**:

$$\mathbf{ln}\, a = (0.693147181) \cdot \mathbf{log_2}\, a = (0.693147181) \cdot \frac{\log a}{0.30103}$$
$$= (2.302\,585\,0) \cdot \mathbf{log}\, a$$

Some important **properties of logarithms** are:

$$\mathbf{log}_b\, a = \frac{\mathbf{log}_{b'}\, a}{\mathbf{log}_{b'}\, B} \quad (b' \text{ is "any basis"})$$

$$\mathbf{log}_b\, b = 1$$

$$\mathbf{log}_b (a \cdot c) = \mathbf{log}_b\, a + \mathbf{log}_b\, c$$

$$\mathbf{log}_b \left(\frac{a}{c}\right) = \mathbf{log}_b\, a - \mathbf{log}_b\, c$$

$$\mathbf{log}_b\, a^n = n \cdot \mathbf{log}_b\, a$$

Chapter 11

The ***ABN method***

The *ABN method* (*Algorithm Based on Numbers*) was devised by *Jaime Martínez Montero*, Doctor of Philosophy and Educational Sciences and Inspector of Education. The flexibility it provides when calculating makes it an ideal complement to this book; this chapter briefly tells you how to perform addition, subtraction, multiplication, and division using this method.

ABN **addition method**

The advantage of this method is that the person who operates can choose not only the number of steps, but also the amounts to add in each of them, so that the sums are more comfortable and manageable.

To calculate any sum, three columns are enough; one for each addend and one to determine the amount to move from one of the addends to the other:

Amount	*Addend 1*		*Addend 2*
	741	+	**4**23
300	**1041**		**123**
103	11**4**4		**2**0
20	**1164**		0

The example above adds 423 to 741 in three steps:

- In the first, 300 is transferred from 423 to 741 (since it is convenient to add 7 and 3 to make 10); 423 minus 300 is **123**; and 741 plus 300 is **1041**.
- In the second, 103 is moved from 123 to 1041; 123 minus 103 is subtracting 1 from 1 and 3 from 3, that is, 0**20**; and 1041 plus 103 is adding 1 to the hundreds and 3 to the units, namely **1144**.
- Finally, 20 is transferred to 1144, which consists of adding 2 to the tens, that is, 1164.

The result of the operation is **1164**.

ABN subtraction **method (*Detraction*)**

It's as flexible as addition. It needs the same columns as before, but this time we need to remove the same amount (the one that is more comfortable for us) from the two operands until one of them is zero.

Three columns allow us to calculate any subtraction; one for the minuend, another for the subtrahend and an additional one to indicate the amount to be subtracted from both the minuend and the subtrahend:

Subtract	*Minuend*		*Subtrahend*
	741	–	**4**23
400	3**41**		**23**
21	3**20**		**2**
2	**318**		0

This subtracts 423 from 741 in three steps:

- In the first, 400 is subtracted from 741 and 423.
- In the second, 21 is subtracted from 341 and 23.
- In the last one, 2 is subtracted from 320 and 2.

The result of the operation is **318**.

ABN subtraction (ASCENDING ladder)

Two columns are required; this time **we add a certain amount** (the most comfortable for us) *to the* ***smallest operand*** until it equals the other operand.

In the largest amount column, we indicate the amounts to be added and in the other each partial sum. It ends when this value is equal to the largest of the operands. The result of the operation is the sum of all the quantities added:

Minuend		*Subtrahend*
741	−	**4**23
300		**723**
7		**730**
11		**741**
+= **318**		

This subtracts 423 from 741 in three steps:

- In the first, we add **300** to 423.
- In the second, we add **7** to 723.
- In the third and last one, we add **11** to 730.

The result of the operation is **318** (300 *plus* 7 *plus* 11).

ABN subtraction (DESCENDING ladder)

Two columns are required; this time **we subtract a certain amount** (the most comfortable for us) *from the operand with the **largest amount*** until it's equal to the other operand.

The amounts to be subtracted are indicated in the largest amount column and each partial subtraction is indicated in the other. It ends when this value is equal to the smallest of the operands. The result of the operation is the sum of all the amounts subtracted; in the example, the result of the operation is **318** (301 *plus* 7 *plus* 10):

Minuend		*Subtrahend*
741	−	423
301		4**40**
7		4**3**3
10		**423**
+= **318**		

where we have subtracted 423 from 741 in three steps:

- In the first, we subtract **301** from 741.
- In the second, we subtract **7** from 440.
- In the third, we subtract **10** from 433.

ABN multiplication **method**

To multiply two numbers, we take one of them and generate a column for each of its constituent parts (units, tens and hundreds, etc.); We proceed in the same way with the other, but this time adding a row for each of its parts (units, tens and hundreds, etc.). Depending on the ability of each one, the partition in terms of the second factor could be done partially (exchanging complexity for fewer rows) or not done (we would have only one row). Later we fill the table thus formed with the product of each element of the corresponding row and column and add them (normally first each row and then all of them):

×	**7000**	**500**	**40**	**6**	*Totals*
40	280 000	20 000	1 600	240	+301 840
8	56 000	4 000	320	48	+60 368
		7546	×	**48**	= **362 208**

The product **7546** × **48** is the sum of the rows of totals (**362 208**).

ABN division method

Dividing two numbers by the *ABN* method consists of subtracting known values less than the dividend using a *scale (extended or synthetic)* created from the divisor to facilitate the selection of each of the partial quotients that added are the result of the operation.

The **extended scale** is created from the product of the divisor (d) by 1, 5, 10, 50, 100, 500, etc. up to cover the magnitude of the dividend. For example, for $d = 34$, the products of 34 times 1, 10, 100 and 1000 would be calculated first, and from them the rest; $34 \times \mathbf{5}$ is 340 divided by 2, that is, **170** (half of 300 —ten times half of 30— plus half of 40); $34 \times \mathbf{50}$ is ten times $34 \times \mathbf{5}$; and $34 \times \mathbf{500}$ is ten times $34 \times \mathbf{50}$:

$$
\begin{array}{rrl}
\times \mathbf{1} = & 34 \times 1 = & \mathbf{34} \\
\times \mathbf{5} = & 34 \times 5 = & \mathbf{170} \\
\times \mathbf{10} = & 34 \times 10 = & \mathbf{340} \\
\times \mathbf{50} = & 34 \times 50 = & \mathbf{1700} \\
\times \mathbf{100} = & 34 \times 100 = & \mathbf{3400} \\
\times \mathbf{500} = & 34 \times 500 = & \mathbf{17000} \\
\times \mathbf{1000} = & 34 \times 1000 = & \mathbf{34000}
\end{array}
$$

The **synthetic scale** is created from the product of the

divisor (d) by 100, 500, 1000, etc. up to cover the magnitude of the dividend. From this scale we obtain two others by dividing by 10: the first, with the products of the divisor by 10, 50, 100, etc. and the second, with the products of the divisor by 1, 5, 10, etc.

To perform the division, three columns are required; the first for the dividends (the initial and the partial ones), the second for the quantities to be subtracted and the last one for each partial quotient and its sum (which is the result of the operation). E.g., the *ABN division* of 50755 *by* 34:

		÷ 34	
Dividend	*Subtract*	*Quotient*	***Synthetic scale***
50755	**34**000	**1000**	(34)
16755	**136**00	400	× 100 = 3400
3155	**306**0	90	× **500** = **17000**
95	**68**	2	× **1000** = 34000
Rest = **27**		+= **1492**	× 1500 = 51000

The comparison of the dividend with the right column of the scale determines the quotient (the value to the left of it or an approximate one). The remainder is the last dividend and the sum of quotients is the quotient.

Bibliography

The Trachtenberg Speed System of Basic Mathematics. *(Translated an adapted by Ann Cutier and Rudolph McShane).* DOUBLEDAY & COMPANY, INC. GARDEN CITY, NEW YORK, 1960. Library of Congress Catalog Card Number 60-13513.

MATEMÁTICAS. Matemáticas modernas *(Biblioteca Hispania Ilustrada.* ***Luis Postigo****).* Editorial Ramón Sopena, S. A. ISBN 84-303-0194-1.

PROBLEMAS RESUELTOS de ANÁLISIS MATEMÁTICO *(Tercera edición.* ***José del Río Sánchez y Leopoldo Suárez Lago****).* *Imprime y distribuye:* Gráficas PAPEL, Plaza de Anaya nº 27. Depósito Legal: S-51-1.987.

Curso de Análisis Matemático I *(Instituto Universitario de Ciencias de la Educación.* ***J. Escuadra Burrieza, J. Rodríguez Lombardero, A. Tocino García****).* Ediciones Universidad de Salamanca, Apartado de correos nº 325. Salamanca. ISBN 84-7800-044-5. Depósito Legal: S. 368 - 1.991.

Química General*.* *TERCERA EDICIÓN (Segunda edición en español).* ***Kenneth W. Whitten, Kenneth D. Gailey, Raymond E. Davis****.* McGraw-Hill. ISBN 968-422-985-2.

WikiABN *(página web de Internet dedicada al método ABN).* La URL a diciembre de 2021 es https://wikiabn.com/

Printed in Great Britain
by Amazon